T0135965

Proceedings

of the

4th International Beilstein Symposium

on

EXPERIMENTAL STANDARD CONDITIONS

OF

ENZYME CHARACTERIZATIONS

September 13th – 16th, 2009

Rüdesheim/Rhein, Germany

Edited by Martin G. Hicks and Carsten Kettner

BEILSTEIN-INSTITUT ZUR FÖRDERUNG DER CHEMISCHEN WISSENSCHAFTEN

Trakehner Str. 7 – 9
60487 Frankfurt
Germany

Telephone: +49 (0)69 7167 3211   **E-Mail:**   info@beilstein-institut.de
Fax: +49 (0)69 7167 3219   **Web-Page:** www.beilstein-institut.de

IMPRESSUM

Experimental Standard Conditions of Enzyme Characterizations, Martin G. Hicks and Carsten Kettner (Eds.), Proceedings of the 4th Beilstein-Institut Symposium, September 13th – 16th 2009, Rüdesheim, Germany.

Bibliographic information published by the *Deutsche Bibliothek*.
The *Deutsche Bibliothek* lists this publication in the *Deutsche Nationalbibliografie*; detailed bibliographic data are available in the Internet at http://dnb.ddb.de.

ISBN 978-3-8325-2606-1

Layout by: Hübner Electronic Publishing GmbH   Printed by   Logos Verlag Berlin GmbH
                    Steinheimer Straße 22a   Comeniushof, Gubener Str. 47
                    65343 Eltville   10243 Berlin
                    www.huebner-ep.de   www.logos-verlag.de
Cover Illustration by: Bosse und Meinhard
                    Kaiserstraße 34
                    53113 Bonn
                    www.bosse-meinhard.de

 **Beilstein-Institut**

PREFACE

The post-"omic" era can be characterized by investigations of dynamic processes within and between cells, tissues and organs. Such investigations are carried out using a combination of interdisciplinary procedures at both the theoretical and experimental level. One aspect of intra-cellular dynamics is the determination of complex metabolic networks and their high dynamic behavior, and their associated mechanistic pathways. Continuous technical and methodological advances and improvements have meant that biochemical pathway analysis can now be carried out in much greater depth and with increased efficiency and accuracy.

Unfortunately, such progress has led to a confusing and highly undesirable situation with respect to trying to make the maximum use of the experimental data derived from the functional characterizations of enzymes since a variety of experimental designs and analytical methods have been employed. The result is that there is a lack of systematic collections of comparable func-tional enzyme data. The pre-requisite for both comparability and reliability of such data is the provision of minimum information about experimental design and experimental results as well as standardization of the conditions and procedures involved in the experiments.

However, the current position is not encouraging: the quality of reported experimental data of enzymes is insufficient for the needs of systems level investigations and thus is, in point of fact, neither applicable for modeling and simulation nor for the functional characterization of the individual cellular components. Consequently, a high quality level balance between experimental in-put data and modeled out-put data needs to be created.

The STRENDA Commission (*St*andards for *R*eporting *En*zyme *Da*ta), founded in 2003, is con-cerned with the improvement of the quality of reporting functional enzyme data to support, *inter alia*, enzyme kinetics for application in the *in silico* investigation of biological systems. The Commission has developed a set of guidelines for the reporting of data in publications. These guidelines (http://www.strenda.org/documents) along with the recommendations of a number of other groups that are also concerned with the standardization of reporting and experimental procedures (http://www.mibbi.org/) are intended to pave the way to *Good Publication Practice* to ensure data quality and data identification.

This 4th ESCEC symposium, organized by the Beilstein-Institut together with the STRENDA Commission, provided a platform to discuss the checklists, to consider further suggestions and to improve the existing recommendations. The STRENDA group took also the opportunity to discuss about the presented STRENDA electronic data capturing tool with members from diverse standardizations initiatives and systems biology groups such as MIBBI, YSBN, EFB and SYS-MO, editorial board members from journals and all participating experimentalists and theoreti-cians. The focus of this presentation was to find answers on the questions such as how to organize and store these massive data sets in standard and easily accessible forms, which new experimental tools have to be developed to gather and configure such data into interactive models, which parameters should be measured, what kind of data constitute the minimum required information, and which experimental conditions should be recommended. The reviewed (provisional) version of this STRENDA data capturing tool can be found at:

https://134.169.106.6/strenda2/index.php?option = com_wrapper&Itemid = 8.

We would like to thank particularly the authors who provided us with written versions of the papers that they presented. Special thanks go to all those involved with the preparation and organization of the symposium, to the chairmen who piloted us successfully through the sessions and to the speakers and participants for their contribution in making this symposium a valuable and fruitful event.

Frankfurt/Main, September 2010   Carsten Kettner
                                                           Martin G. Hicks

V

Beilstein-Institut

Experimental Standard Conditions of Enzyme Characterizations,
September 13th – 16th, 2009, Rüdesheim/Rhein, Germany

CONTENTS

Page

Beilstein-Institut

Experimental Standard Conditions of Enzyme Characterizations,
September 13th – 16th, 2009, Rüdesheim/Rhein, Germany

VI

Page

Experimental Standard Conditions of Enzyme Characterizations,
September 13th – 16th, 2009, Rüdesheim/Rhein, Germany

Structure, Function and Evolution of Fosfomycin Resistance Proteins in the Vicinal Oxygen Chelate Superfamily

Richard N. Armstrong[*], Paul D. Cook and Daniel W. Brown

Departments of Biochemistry and Chemistry, Center in Molecular Toxicology, and the Vanderbilt Institute of Chemical Biology, Vanderbilt University, Nashville, TN, 37232–0146, U.S.A.

E-Mail: [*]r.armstrong@vanderbilt.edu

Received: 19th March 2010 / Published: 14th September 2010

Abstract

The Vicinal Oxygen Chelate (VOC) superfamily embodies a function-ally diverse set of enzymes that catalyze both acid-base and electron transfer chemistries [1]. A subset of these enzymes is known to confer microbial resistance to the antibiotic fosfomycin by three different mechanisms. The resistance proteins FosA (a glutathione transferase), FosB (a thiol transferase) and FosX (an epoxide hydrolase) are found in both Gram-negative and Gram-positive pathogenic microorganisms. These proteins have been proposed to be evolutionarily related to a catalytically promiscuous progenitor (FosX$_{Ml}$) encoded in a *phn* oper-on in *Mesorhizobium loti* [2]. We recently reported that more robust FosA activity could be evolved by homologous recombination experi-ments with a FosA gene and the gene encoding the promiscuous FosX$_{Ml}$ [3]. This report is incorrect. The "evolved" proteins that were characterized appear not to be the result of homologous recombination but rather due to random mutations in a mutant gene that contaminated the original recombination experiments. This paper first summarizes what is known about the evolutionary relationships among these pro-teins and then points to new lines of investigation, particularly with respect to Gram-positive microorganisms.

INTRODUCTION

Fosfomycin, *(1R,2S)*-epoxypropylphosphonic acid, **1**, is a natural product that has potent, broad-spectrum antimicrobial activity against both Gram-positive and Gram-negative micro-organisms [4, 5]. Fosfomycin acts by the covalent inactivation of the enzyme UDP-N-acetylglucosamine-3-enolpyruvyltransferase or MurA. The MurA enzyme catalyzes the first committed step in peptidoglycan biosynthesis, the addition of pyruvate to the 3'-hydroxyl group of UDP-N-acetylglucosamine (UDP-GlucNAc), providing the three-carbon unit that ultimately links the glycan copolymer and the peptide units of the cell wall. The inactivation of MurA occurs only in the presence of UDP-GlucNAc and involves the alkylation of an active site cysteine residue (C115 in the *E. coli* enzyme) Scheme 1 [6, 7]. The structure of the covalently inhibited enzyme has been reported in [8].

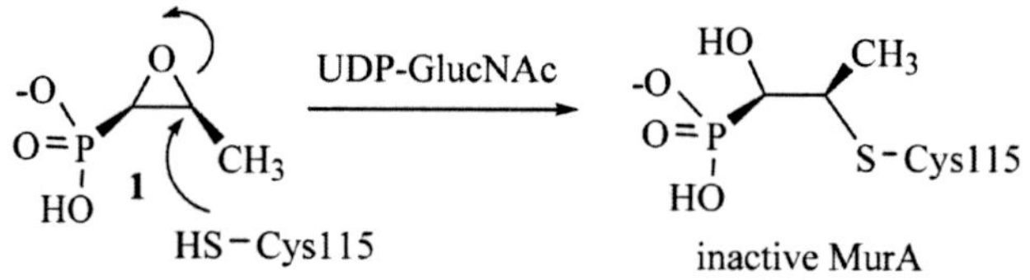

Scheme 1. Inactivation of MurA by fosfomycin.

A decade after the introduction of fosfomycin into the clinic, a plasmid-mediated resistance to the antibiotic was observed in clinical isolates obtained from patients treated with the drug [9 – 11]. Additional investigations indicated that the resistance gene encoded a 16 kDa polypeptide that catalyzed the addition of glutathione (GSH) to the antibiotic, rendering it inactive [12, 13].

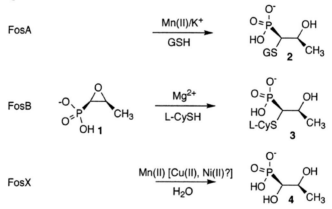

Figure 1. Reactions catalyzed by the fosfomycin resistance proteins FosA, FosB and FosX. The FosA catalyzed reaction with GSH requires the monovalent cation (K$^+$) for optimal activity. The FosB reaction is optimal with L-cysteine as the nucleophile to form adduct **3** but may use alternative nucleophiles not yet characterized.

The fosfomycin hydrolase enzymes (FosX) appear to use Mn(II) in most instances, but with some enzymes, Cu(II) and Ni(II) also work as well as, if not better than, Mn(II). The diol, **4**, is formed by addition of water at C 1.

In the period between 1996 and 1999, we established that this resistance protein (now termed FosA) was an Mn(II) and K^+-dependent GSH transferase that catalyzed the formation of the GSH adduct (**2**) at C 1 of the antibiotic, as illustrated in Figure 1 [14, 15]. Moreover, we discovered that there are additional mechanisms of resistance catalyzed by enzymes in the same superfamily [2, 16 – 18].

An analysis of primary sequence information and available three-dimensional structural data indicates that currently known fosfomycin resistance proteins are evolutionarily related and fall into three basic categories as illustrated in Figures 1 and 2. Although the resistance protein identified was encoded on a multi-drug resistance plasmid, numerous others have been identified encoded in the genomes of human pathogens.

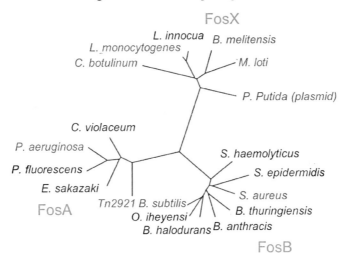

Figure 2. Sequenced-based segregation of the subfamilies of the fosfomycin resistance proteins FosA, FosB and FosX. The three subfamilies are shown in green. The gene products shown in red have been expressed and characterized biochemically, and the crystal structures of the proteins have been determined. The gene products shown in blue have been expressed and characterized biochemically.

MATERIALS AND METHODS

The materials and methods for this work have been reported previously [3, 17] and will not be reiterated here.

FUNCTIONAL AND EVOLUTIONARY RELATIONSHIP BETWEEN FOSA AND FOSX

The FosA and FosX proteins share about 30–35% sequence identity. The FosA proteins that have been characterized do not catalyze the hydration of fosfomycin, and the highly evolved FosX enzymes such as those from *Listeria monocytogenes* and *Pseudomonas putida* do not catalyze the addition of GSH to fosfomycin. However, the first protein to be discovered that catalyzed the hydration of fosfomycin was also found to catalyze the addition of GSH to the antibiotic. This protein (FosX$_{Ml}$) is encoded by a gene in an operon in *Mesorhizobium loti* that probably functions in phosphonate catabolism. The FosX$_{Ml}$ enzyme is not a good resistance protein but could be an evolutionary progenitor of genuine resistance proteins that have FosA, FosB or FosX activities. To examine this possibility we set out to evolve a robust FosA enzyme from the gene encoding FosX$_{Ml}$ by either rational, structure-based mutagenesis or homologous recombination.

The rational design strategy was guided by structural comparisons of the FosA and FosX proteins and computational prediction of the GSH binding site in FosA. The principal differences in or near the active sites of the two proteins include three residues near the metal center and a loop located adjacent to the active site that serves to help bind GSH and the K$^+$-ion in FosA (Figure 3). In a recent report we described the construction of a triple mutant (E44G/F46Y/M56S) in the FosX$_{Ml}$ protein [3]. The FosX$_{Ml}$(triple mutant) lost all fosfomycin hydrolase activity and gained considerable GSH transferase activity. Replacement of the loop region of the triple mutant of FosX$_{Ml}$ with that of the FosA from *Pseudomonas aeruginosa* (FosA$_{Pa}$) resulted in a FosX$_{Ml}$(triple+loop mutant) and a very large loss of GSH transferase catalytic activity (unpublished results).

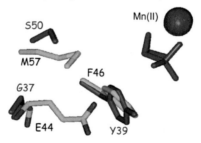

```
FosA_Pa    87-REWKQNR---SEGDSFYFL-103
FosX_Ml    92-DMRPPRPRVEGEGRSIYFY-110
```

Figure 3. (Top) Overlay of the FosA$_{Pa}$ (blue) and FosX$_{Ml}$ structures illustrating the residues near the substrate also shown in blue. The Mn(II) ions are shown as blue and purple spheres. (Bottom) Sequence alignment of the loop region of FosA$_{Pa}$ and FosX$_{Ml}$. Residues highlighted in red are involved in GSH binding, those in green are K$^+$-binding residues and those in blue are the insert unique to FosX.

In the same paper [3] we also reported what appeared to be successful homologous recombination (gene shuffling) experiments between the FosX$_{MI}$ and FosA$_{Pa}$ genes resulting in six different recombinant proteins [3]. Two of the proteins were characterized in detail both with respect to their steady-state kinetics and their ability to confer resistance to fosfomycin in *E. coli*. In a subsequent effort to extend these studies, we attempted to reproduce the initial experiments using freshly prepared and sequenced DNA samples for FosX$_{MI}$ and FosA$_{Pa}$. After numerous additional experiments we found that we were unable to reproduce the original results. This failure prompted a detailed analysis of the nucleotide sequences of the six recombinants. The sequence analysis shows that the recombinants were not the result of homologous recombination but strongly suggest that they were due to contamination of one of the original DNA samples (most likely the FosX$_{MI}$ plasmid) with the plasmid encoding the FosX$_{MI}$(triple+loop mutant). A correction to this effect is being prepared for publication in *Biochemistry*.

The six variants isolated in the antibiotic screen most likely arose either from "self recombination" of the FosX$_{MI}$(triple+loop mutant) or recombination between fragments of the FosX$_{MI}$ gene and those of the FosX$_{MI}$(triple+loop mutant). The gene shuffling process requires the use of the polymerase chain reaction (PCR), which is prone to errors in base incorporation. The six variants that were isolated are probably due to errors in the PCR steps of the homologous recombination process. That said, it is important to indicate that we have no direct evidence that this is the case.

Although it is clear that the FosA/B/X proteins are evolutionarily related based on sequence and structure, it is less obvious that the FosX$_{MI}$ protein is a promiscuous progenitor of true fosfomycin resistance proteins. It is clearly a promiscuous enzyme, being able to use both water and GSH as the nucleophile. However, any direct evolutionary connection remains experimentally obscure, perhaps due to evolutionary drift among the genes.

What is the Thiol Substrate for FosB?

In contrast to the FosA and FosX proteins, the FosB resistance enzymes have not been fully characterized. For example, there is no three-dimensional structure of a FosB from any source. One of the most important unanswered questions with respect to the FosB proteins is the identity of the thiol substrate. Gram-positive microorganisms such as *Bacillus subtilis* and *Staphylococcus aureus* do not make glutathione suggesting that it is not the native thiol substrate for FosB. We originally reported that L-cysteine was the most likely candidate substrate for the FosB in *B. subtilis* based on its relatively high concentration (0.1 mM) in the organism. The enzyme does catalyze the reaction but the efficiency of the reaction is not very impressive with a k_{cat}/K_M^{fos} of about 10^3 M^{-1}s^{-1} and a $k_{cat}=5$ s^{-1}.

A very recent paper [19] reported a newly discovered thiol that is a derivative of L-cysteine. The compound is found in significant concentrations in many Gram-positive microorganisms, particularly, in strains of *Bacillus*. The concentrations are comparable to or greater than that of L-cysteine. Although the function of this newly discovered molecule is not known, it may well the natural substrate for FosB. This remains a very fertile area of investigation.

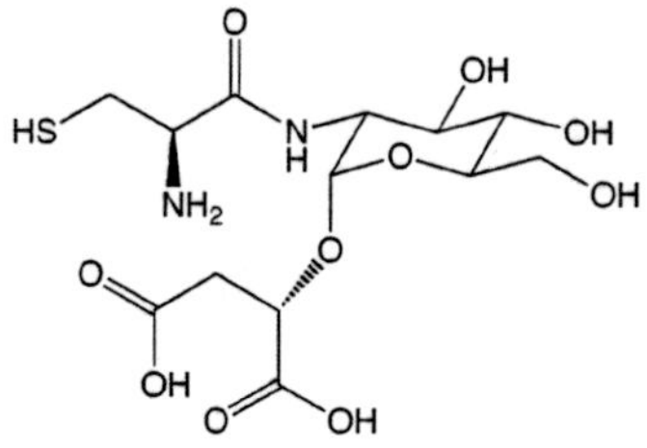

Figure 4. Chemical structure of bacillithiol.

ACKNOWLEDGEMENT

This work was supported by National Institutes of Health Grants R56 AI042756, T32 ES 007028, P30 ES 000267.

REFERENCES

[1]   Armstrong, R.N. (2000) Mechanistic Diversity in a Metalloenzyme Superfamily. *Biochemistry* **39**:13625 – 13632.
       doi: http://dx.doi.org/10.1021/bi001814v.

[2]   Fillgrove, K.L., Pakhomova, S., Newcomer, M.E., Armstrong, R.N. (2003) Mechanistic diversity of fosfomycin resistance in pathogenic microorganisms. *J. Am. Chem. Soc.* **125**:15730 – 15731.
       doi: http://dx.doi.org/10.1021/ja039307z.

[3]   Brown, D.W., Schaab, M.R., Birmingham, W.R., Armstrong, R.N. (2009) Evolution of the Antibiotic Resistance Protein, FosA, is linked to a Catalytically Promiscuous Progenitor. *Biochemistry* **48**:1847 – 1849.
       doi: http://dx.doi.org/10.1021/bi900078q.

[4]   Hendlin, D., Stapley, E.O., Jackson, M., Wallick, H., Miller, A.K., Wolf, F.J., Miller, T.W., Chaiet, L., Kahan, F.M., Flotz, E.L., Woodruff, H.B., Mata, J.M., Hernandez, S., Mochales, S. (1969) Phosphonomycin, a new antibiotic produced by strains of Streptomyces. *Science* **166**:122 – 123.
       doi: http://dx.doi.org/10.1126/science.166.3901.122.

[5]   Christensen, B.G., Leanza, W.J., Beattie, T.R., Patchett, A.A., Arison, B.H., Ormond, R.E., Kuehl, F.A., Albers-Schonberg, G., Jardetzky, O. (1969) Phosphonomycin: structure and synthesis. *Science* **166**:123 – 124.
       doi: http://dx.doi.org/10.1126/science.166.3901.123.

[6]   Kahan, F.M., Kahan, J.S., Cassidy, P.J., Kroop, H. (1974) The mechanism of action of fosfomycin (phosphonomycin). *Ann. N. Y. Acad Sci.* **235**:364 – 385.
       doi: http://dx.doi.org/10.1111/j.1749-6632.1974.tb43277.x.

[7]   Marquardt, J.L., Brown, E.D., Lane, W.S., Haley, T.M., Ichskawa, Y., Wong, C.-H., Walsh, C.T. (1994) Kinetics, stoichiometry, and identification of the reactive thiolate in the inactivation of UDP-GlcNAc enolpyruvoyl transferase by the antibiotic fosfomycin. *Biochemistry* **33**:10646 – 10651.
       doi: http://dx.doi.org/10.1021/bi00201a011.

[8]   Skarzynski, T., Mistry, A. Wonacott, A., Hutchinson, S.E. Kelly, V.A., Duncan, K. (1996) Structure of UDP-N-acetylglucosamine enolpyruvyl transferase, an enzyme essential for the synthesis of bacterial peptidoglycan, complexed with substrate UDP-N-acetylglucosamine and the drug fosfomycin. *Structure* **4**:1465 – 1474.
       doi: http://dx.doi.org/10.1016/S0969-2126%2896%2900153O.

[9]   Mendoza, M.C., Garcia, J.M., Llaneza, J., Mendez, J.F., Hardisson, C., Ortiz, J.M. (1980) Plasmid determined resistance to fosfomycin in *Serratia marcescens*. *Antimicrob. Agents Chemother.* **18**:215 – 219.

[10] Llaneza, J., Villar, C.J., Salas, J.A., Suarez, J.E., Mendoza, M.C., Hardisson, C. (1985) Plasmid-mediated fosfomycin resistance is due to enzymatic modification of the antibiotic. *Antimicrob. Agents Chemother.* **28**:163–164.

[11] Villar, C.J., Hardisson, C., Suarez, J.E. (1986) Cloning and molecular epidemiology of plasmid-determined fosfomycin resistance. *Antimicrob. Agents Chemother.* **29**:309–314.

[12] Arca, P., Rico, M., Brana, A.F., Villar, C.J., Hardisson, C., Suarez, J.E. (1988) Formation of an adduct between fosfomycin and glutathione: a new mechanism of antibiotic resistance in bacteria. *Antimicrob. Agents Chemother.* **32**:1552–1556.

[13] Arca, P., Hardisson, C., Suarez, J.E. (1990) Purification of a glutathione S-transferase that mediates fosfomycin resistance in bacteria. *Antimicrob. Agents Chemother.* **34**:844–848.

[14] Bernat, B.A., Laughlin, L.T., Armstrong, R.N. (1997) Fosfomycin resistance protein (FosA) is a manganese metalloglutathione transferase related to glyoxalase I and the extradiol dioxygenases. *Biochemistry* **36**:3050–3055.
doi: http://dx.doi.org/10.1021/bi963172a.

[15] Bernat, B.A., Laughlin, L.T., Armstrong, R.N. (1999) Elucidation of a monovalent cation dependence and characterization of the divalent cation binding site of the fosfomycin resistance protein, FosA, *Biochemistry* **38**:7462–7469.
doi: http://dx.doi.org/10.1021/bi990391y.

[16] Cao, M., Bernat, B.A., Wang, Z., Armstrong, R.N., Helmann, J.D. (2001) FosB, a cysteine-dependent fosfomycin resistance protein under the control of σ^W, an extracytoplasmic function σ factor in *Bacillus subtilis*. *J. Bacteriol.* **183**:2380–2383.
doi: http://dx.doi.org/10.1128/JB.183.7.2380-2383.2001.

[17] Rigsby, R.E., Fillgrove, K.L., Beihoffer, L., Armstrong, R.N. (2005) Fosfomycin resistance proteins: A nexus of glutathione transferases and epoxide hydrolases in a metalloenzyme superfamily. *Methods Enzymol.* **401**:367–379.
doi: http://dx.doi.org/10.1016/S0076-6879%2805%2901023-2.

[18] Fillgrove, K.L., Pakhomova, S., Schaab, M.R., Newcomer, M.E., Armstrong, R.N. (2007) Structure and mechanism of the genomically encoded fosfomycin resistance protein, FosX, from *Listeria monocytogenese*. *Biochemistry* **46**:8110–8120.
doi: http://dx.doi.org/10.1021/bi700625p.

[19] Newton, G.I., Rawat, M., La Clair, J.J., Jothivasan, V.K., Budiarto, T., Hamilton, C.J., Claibourne, A., Helmann, J.D., Fahey, R.C. (2009) Bacillithiol is an antioxidant thiol produced in Bacilli. *Nature Chem. Biol.* **5**:625–627.
doi: http://dx.doi.org/10.1038/nchembio.189.

 Beilstein-Institut

Experimental Standard Conditions of Enzyme Characterizations,
September 13th – 16th, 2009, Rüdesheim/Rhein, Germany

FUNCTIONAL ANNOTATION OF ORPHAN ENZYMES WITHIN THE AMIDOHYDROLASE SUPERFAMILY

FRANK M. RAUSHEL

Department of Chemistry, Texas A&M University,
College Station, TX 77843, U.S.A.

E-Mail: raushel@tamu.edu

Received: 23rd December 2009 / Published: 14th September 2010

ABSTRACT

The elucidation of the substrate profiles for enzymes of unknown function is a difficult and demanding problem. A general approach to this problem combines bioinformatics and operon context, computational docking to X-ray crystal structures, and the utilization of focused chemical libraries. These methods have been applied to the identification of novel substrates for enzymes of unknown function within the amidohydrolase superfamily. Operon context and X-ray crystallography was utilized in the identification of *N*-formimino-L-glutamate as the substrate for Pa5105 from *Pseudomonas aeruginosa* and D-galacturonate for Bh0493 from *Bacillus halodurans*. Focused substrate libraries were used to identify *N*-acetyl-D-glutamate as the substrate for Bb3285 from *Bordetella bronchiseptica* and L-Xaa-L-Arg/Lys as the substrate for Cc2672 from *Caulobacter crescentus*. Computational docking of potential high energy intermediates was used to determine that Tm0936 from *Thermotoga maritima* catalyzed the deamination of *S*-adenosyl homocysteine.

INTRODUCTION

The recent advent of high throughput DNA sequencing efforts has significantly enhanced the number of completely sequenced bacterial genomes. The number of non-redundant genes that have been deposited in the public databases now exceeds 8 million entries. A close examination of these sequences indicates that a significant fraction of the proteins and enzymes coded by these gene sequences have an unknown, uncertain or incorrect functional annotation. This fact suggests that there are a substantial number of biochemical reactions that remain to be discovered. However, annotating enzymes of unknown function, based upon the protein sequences alone, is a difficult and demanding problem [1]. One strategy toward the solution to this problem utilizes a combination of bioinformatics, computational docking to X-ray structures or homology models, and library screening. Our efforts in this area have focused on the annotation of function for members of the amidohydrolase super-family.

The amidohydrolase superfamily (AHS) was first identified in 1997 by Sander and Holm who recognized the structural similarities among urease, phosphotriesterase, and adenosine deaminase [2]. All three proteins fold as a distorted $(\beta/\alpha)_8$-barrel structure and possess either a binuclear or mononuclear metal centre within the active site [3]. Most of the experimentally characterized members of the AHS have been shown to catalyze the hydrolysis of amide and ester substrates contained within carbohydrates, peptides, and nucleic acids [3]. However, other members of the AHS have been shown to catalyze isomerization, hydration, and decarboxylation reactions. The metal centres in these proteins function to activate solvent water for nucleophilic attack and/or to enhance the reactivity of the substrate [4]. More than 12,000 unique protein sequences in the first 1,000 completely sequenced bacterial genomes that have been deposited in the NCBI have been identified as being part of the AHS. These sequences have been subclassified into 24 clusters of orthologous groups (COGs).

REPRESENTATIVE EXAMPLES

Pa5106: N-Formimino-L-Glutamate Deiminase

One of the first examples for the elucidation of a previously unrecognized function for a member of the amidohydrolase superfamily occurred with Pa5106 [5]. This protein from *Pseudomonas aeruginosa* PA01 is a member of cog0402 and was misannotated as a "probable chlorohydrolase or cytosine deaminase". At the time of this investigation all of the functionally characterized members of cog0402 catalyzed the deamination of aromatic bases and these proteins included guanine deaminase and cytosine deaminase. The hallmark for this COG is an HxxE motif that is found at the end of beta-strand 5 in the $(\beta/\alpha)_8$-barrel structure. In this motif, the histidine residue coordinates to the single divalent cation in the active site while the glutamate functions to shuttle a proton from the hydrolytic water molecule to the deaminated products (xanthine and uracil from guanine and cytosine, respectively).

Examination of the genomic context for Pa5106 within *P. aeruginosa* revealed that this gene was adjacent to a cluster of genes that are known to be involved in the degradation of histidine to glutamate. The histidine degradation pathway (hut operon) is shown in Scheme 1. In this pathway histidine is first deaminated to urocanate by histidine ammonia lyase (HutH) and then urocanase (HutU) converts this product to imidazolone-4-propionate. Imidazolone propionate amidohydrolase (HutI) catalyzes the hydrolysis of imidazolone-4-propionate to *N*-formimino-L-glutamate. In the last step the formimino group of *N*-formimino-L-glutamate is transferred to either tetrahydrofolate or to water by HutG to make the final product, L-glutamate. At first the localization of the gene for Pa5106 next to the hut operon was quite confusing (at least for us) since all four of the known proteins for the conversion of histidine to glutamate were accounted for in this pathway (HutH, HutU, HutI, and HutG) and there was no obvious need for an enzyme that we initially thought would catalyze the deamination of an aromatic base. However, it soon occurred to us that the formimino functional group of *N*-formimino-L-glutamate looked very much like that portion of guanine or cytosine that was deaminated by other members of cog0402 within the AHS. Therefore, we predicted that Pa5106 would catalyze the deimination of *N*-formimino-L-glutamate to *N*-formyl-L-glutamate (HutF) and that the protein designated as HutG would actually catalyze the hydrolysis of *N*-formyl-L-glutamate to formate and L-glutamate [5].

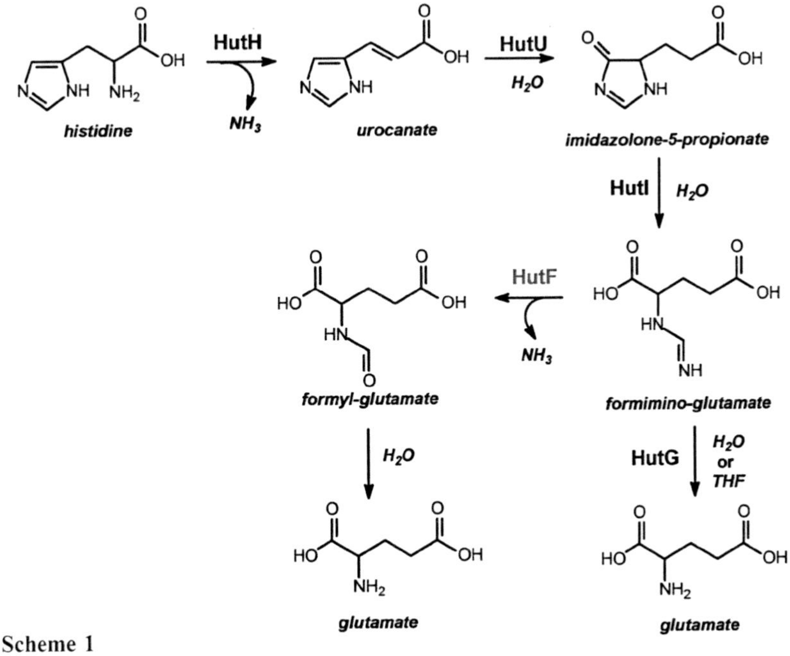

Scheme 1

These predictions proved to be correct. Pa5106 was found to catalyze the deimination of N-formimino-L-glutamate with values for k_{cat}, K_m, and k_{cat}/K_m of 13 s^{-1}, 0.22 mM, and 6×10^4 $M^{-1}s^{-1}$, respectively. The protein originally annotated as HutG (Pa5091) was found to catalyze the hydrolysis of N-formyl-L-glutamate to formic acid and L-glutamate with values of k_{cat}, K_m, and k_{cat}/K_m of 1 s^{-1}, 3.3 mM, and 3×10^2 $M^{-1}s^{-1}$, respectively. These reactions were first discovered over 50 years ago by Tabor and Mehler [6].

Bh0493: D-Galacturonate Isomerase

One of the most diverged members of the AHS to be functionally characterized is Bh0493 from *Bacillus halodurans* [7]. When this protein was first interrogated the sequence identity to any other member of the amidohydrolase superfamily was less than 20% and there was significant doubt that this protein was even a member of the AHS. Nevertheless, the closest structurally characterized homologue that could be identified was uronate isomerase (Tm0064) from *Thermotoga maritima* [PDB code: 1j5 s]. Bh0493 contains a conserved WWF motif at the end of beta-strand 7 that is conserved in all of the known bacterial uronate isomerases but a conserved histidine at the end of beta-strand 5 is missing. Adding to the confusion about the functional identity of Bh0493 as a putative uronate isomerase is the presence of another protein in the genome of *B. halodurans* that is annotated as an uronate isomerase (Bh0705).

The transformations utilized by many bacteria for the metabolism of D-glucuronate and D-galacturonic are shown in Scheme 2. These two uronic acids are isomerised to D-fructuronate and D-tagaturonate, respectively, by a single enzyme, uronate isomerase. In *E. coli* D-fructuronate is subsequently converted to 2-keto-3-deoxy-D-gluconate (KDG) by the combined actions of UxuB and UxuA, whereas D-tagaturonate is transformed to KDG by a different pair of enzymes, UxaB and UxaA. Examination of the operon context for Bh0493 in *B. halodurans* proved informative since the gene for this enzyme was found to be adjacent to two genes homologous to UxaA (Bh0490) and UxaB (Bh0492) whereas the more prototypical uronate isomerase (Bh0705) was adjacent to UxuA (Bh0706) and UxuB (Bh0707). It was therefore postulated for *B. halodurans* that separate isomerases were utilized for the metabolism of D-glucuronate and D-galacturonate [7]. For Bh0705 the value of k_{cat}/K_m for D-glucuronate was determined to be two orders of magnitude greater than for D-galacturonate. For Bh0493 the values of k_{cat}/K_m for the two compounds are essentially the same. The operon contexts and the kinetic constants for the two enzymes capable of isomerising uronic acids in *B. halodurans* are consistent with the primary function for Bh0493 as a D-galacturonate isomerase.

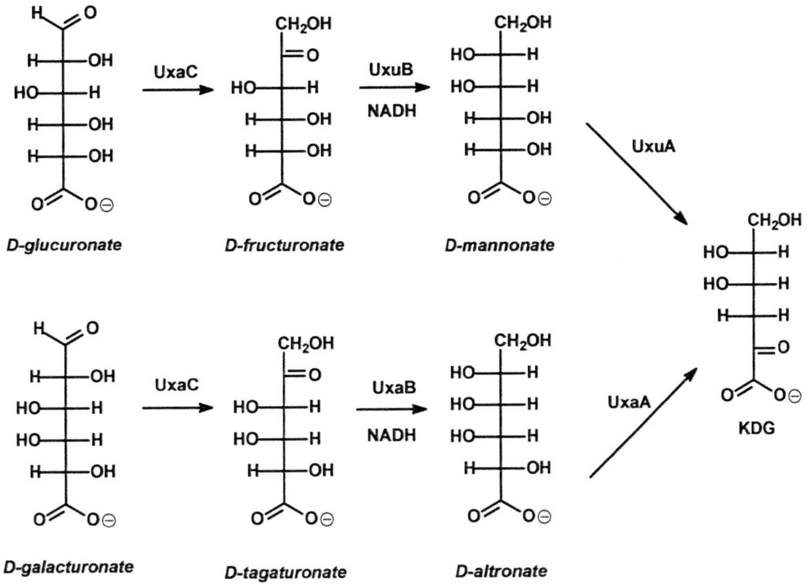

Scheme 2

Bb3285: N-Acetyl-D-Glutamate Deacetylase

The function of Bb3285 from *Bordetella bronchiseptica* was determined primarily through the utilization of a focused chemical library screen since the operon context was of little use in the assignment of function [8]. This protein is found in cog3653 and some of the enzymes in this COG have been functionally annotated as deacetylases or peptidases. Therefore, small substrate libraries containing nearly all possible combinations of L-Xaa-L-Xaa, L-Xaa-D-Xaa, D-Xaa-L-Xaa, N-acyl-D-Xaa, and N-acyl-L-Xaa were tested as substrates for Bb3285 and the products of the hydrolysis reactions quantified by amino acid analysis. Of the substrate libraries tested, the only one that showed any significant formation of a free amino acid after the addition of enzyme was N-acetyl-D-Xaa. This library contained the twenty common amino acids derivatized with an N-acetyl group. The chromatogram from the HPLC analysis is presented in Figure 1. The only substrate for Bb3258 in this library is N-acetyl-D-gluta-mate. The kinetic constants for k_{cat}, K_m, and k_{cat}/K_m were found to be 460 s^{-1}, 88 μM, and 5 x 10^6 M^{-1}s^{-1}, respectively [8].

The identification of N-acetyl-D-glutamate as the primary substrate for Bb3285 enabled us to design an analogue that proved to be a potent inhibitor of this enzyme. The N-methyl phosphonate derivative of D-glutamate resembles the tetrahedral intermediate that would be formed during substrate hydrolysis. The structure of this compound is presented in Scheme 3. This compound is a competitive inhibitor versus N-acetyl-D-glutamate with a K_i value of 460 pM!

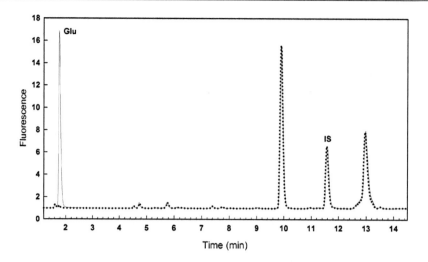

Figure 1. Chromatogram of the *N*-acetyl-D-Xaa library treated with no enzyme (black dots) and 20 nM Bb3285 (red line) for 1 hour at 30 °C. The OPA-derivatized D-glutamate was detected at a retention time of 1.7 minutes in the sample treated with Bb3285. The internal standard is labeled as IS.

In collaboration with the group of Steve Almo at the Einstein College of Medicine we were able to crystallize Bb3285 in the presence of this inhibitor and the molecular interactions are shown in Figure 2 (pbd code: 3giq). In this structure, the two phosphonate oxygens bridge the two metal ions in the binuclear metal centre. The α-carboxylate of the substrate is ion-paired with Lys-250, Arg-376, and Tyr-282. The recognition of the side-chain carboxylate is an ion-pair interaction with Arg-295. The structure of this complex has enabled us to identify another cluster of enzymes within cog3653 that specifically hydrolyze *N*-acyl-D-Hydrophobic amino acid derivatives. The specific examples included Gox1177 from *Gluconobacter oxydans* and Sco4986 from *Streptomyces coelicolor* [8].

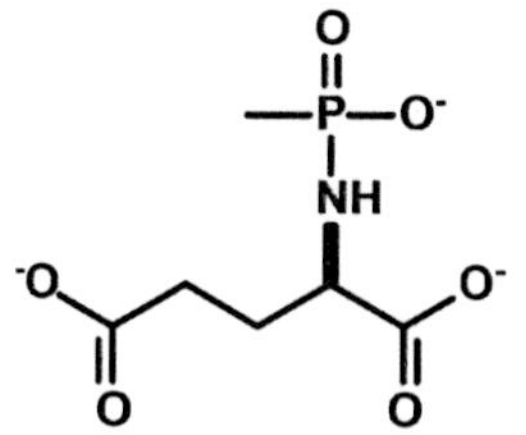

Scheme 3

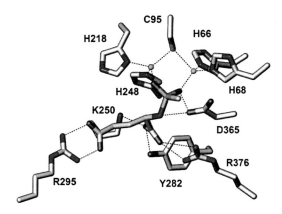

Figure 2. Binuclear Zn (green spheres) active site of Bb3285 with bound inhibitor (pink carbons, orange phosphorus). Enzyme-substrate contacts within 2.0 – 3.5 Å are indicated by dashed lines.

Cc2672: L-*Xaa*-L-*Arg/Lys Dipeptidase*

The substrate profile for Cc2672 from *Caulobacter crescentus* CB15 was determined using a combination of library screening and X-ray crystallography [9]. At the start of this investigation Cc2672 was annotated in the NCBI as a L-Xaa-L-Pro dipeptidase. Similar to the situation with Bb3285 discussed above, the operon context for Cc2672 was of no help in the search for the reaction catalyzed by this enzyme. Therefore, the initial test of catalytic activity employed a broad set of dipeptide libraries that covered most of the nearly 1600 combinations of the twenty common amino acids in the D- and L-configurations. Of the various dipeptide libraries tested, the only ones that displayed significant catalytic turnover with Cc2672 were of the type L-Zaa-L-Xaa. In these substrate libraries a fixed amino acid was placed at the N-terminus (L-Zaa) with a combination of the 20 common amino acids at the C-terminus (L-Xaa). The liberation of free amino acids was monitored as a function of time with ninhydrin as a preliminary measure of the number of dipeptides in the library that serve as substrates. Quantitative amino acid analysis was utilized with a single dipeptide library (for example, L-Ala-L-Xaa) to determine the specific dipeptides that are hydrolyzed and the relative rates of hydrolysis for all of the dipeptides contained within a given dipeptide library. When this was conducted with Cc2672 the only free amino acids detected were L-lysine and L-arginine (and the fixed amino acid at the N-terminus). The substrate specificity for Cc2672 is therefore L-Zaa-L-Arg/Lys. There was very little discrimination among the twenty common amino acids at the N-terminus but an absolute requirement for either arginine or lysine at the C-terminus.

Thus far, attempts to determine the X-ray crystal structure of Cc2672 have failed. However, we have been able, in collaboration with the group of Steve Almo at the Einstein College of Medicine, to determine the structure of a close homologue of this enzyme. The homologue, designated as Sgx9359b (gi|44368820), is a protein whose DNA was originally isolated from the Sargasso Sea. The crystal structure was solved to a resolution of 2.3 Å and two divalent cations were found in the active site (pdb code: 3be7 and 3dug). In one of the subunits arginine is found bound as a product in the active site and thus this structure illustrated how the terminal carboxylate of dipeptide substrates is recognized and reveals the structural determinants for the C-terminal substrate specificity. The α-carboxylate is ion paired with a histidine (His-225) that is found at the end of β-strand 5 and the guanidino group is ion paired with a glutamate (Glu-289) that is found in the loop after β-strand 7. These interactions are illustrated in Figure 3. The structural determinants for substrate specificity have helped to identify the substrate profiles for other members of this superfamily that are specific for the hydrolysis of L-Xaa-L-Hydrophobic dipeptides [10] and dipeptides that terminate in proline [11].

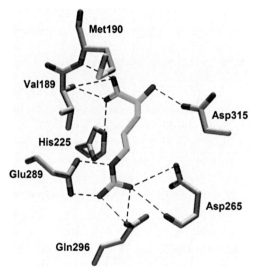

Figure 3. Structure of the active site of Sgx9359b showing the interactions of the product arginine with various residues. Taken from pdb code: 3dug.

Tm0936: S-Adenosyl Homocysteine Deaminase

The identification of the catalytic function for Tm0936 from *Thermotoga maritima* was accomplished largely through the utilization of computational docking to an existing X-ray crystal structure [12]. These calculations were done in collaboration with the group of Brian Shoichet at UC-San Francisco. The X-ray structure of Tm0936 showed that this enzyme contained a single zinc in the active site and the conserved HxxE motif at the end of β-strand 5 placed this enzyme within cog0402 (pdb code: 1plm and 1j6 p). This enzyme is in the same COG as Pa5106, cytosine deaminase and guanine deaminase, and thus it was highly likely that the overall reaction would involve a deamination reaction. For the docking calculations, the Shoichet laboratory created small molecule mimics that resembled the putative transition state intermediates for the hydrolysis reactions. The strategy here was based on the assumption that molecules that resembled the transition states would be more selective for the active site than simple ground state molecules [13]. The obvious complication for these types of docking calculations is the high potential for conformational changes in the protein structure upon binding of the substrate to the active site. Nevertheless, the entire KEGG library of compounds that possess a hydrolytic site was computationally docked into the active site of Tm0936. Of the top 100 hits, nearly 40% of the compounds were modifications of adenosine. This result provided a high degree of confidence that an adenosine derivative would be deaminated by Tm0936. Of the compounds tested, the best substrate was S-adenosyl homocysteine (SAH) followed by thiomethyl adenosine (TMA) and adenosine itself. The values of k_{cat}/K_m for the three compounds were greater than 10^5 $M^{-1}s^{-1}$. As a test of the correctness of the proposed docking pose relative to the binding of actual ligands to the active site we enzymatically prepared the product of the reaction by incubating SAH with Tm0936 and then isolated the product, S-inosyl homocysteine (SIH). The SIH was then used in a co-crystallization of Tm0936. The structural overlay between the docking pose of the proposed high energy intermediate and the conformation of SIH bound in the active site of Tm0936, determined by X-ray diffraction methods, was excellent (pdb code: 2plm). The conversion of SAH to SIH had not previously been recognized as a metabolic transformation and it is shown in Scheme 4.

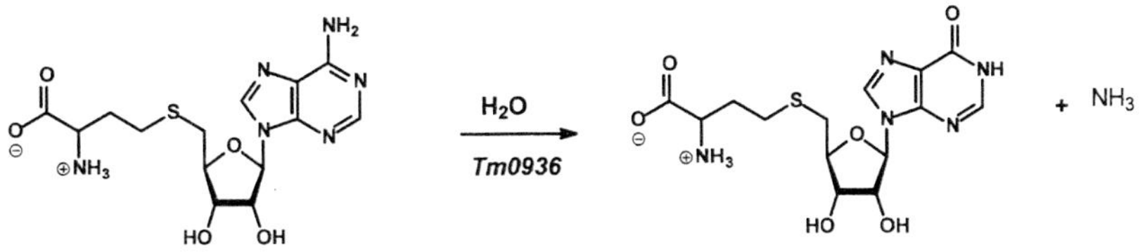

Scheme 4

CONCLUSIONS

A significant fraction of the genes contained within recently sequenced bacterial genomes have an unknown function. We have attempted to develop a broad-based strategy for determining the function of enzymes belonging to the amidohydrolase superfamily. Employment of bioinformatics and operon context, computational docking of intermediates and substrates to active sites, and screening with focused chemical libraries have enabled the identification of novel substrates for a variety of enzymes of unknown function.

ACKNOWLEDGMENT

The work described in this paper was supported in part by the National Institutes of Health (GM 71790) and the Robert A. Welch Foundation (A-840).

REFERENCES

[1]   Gerlt, J.A. and Babbitt, P.C. (2000) Can sequence determine function? *Genome Biology* **5**:1 – 10.
doi: http://dx.doi.org/10.1186/gb-2000-1-5-reviews0005.

[2]   Holm, L. and Sander, C. (1997) An evolutionary treasure: unification of a broad set of amidohydrolases related to urease. *Proteins* **28**:72 – 82.
doi: http://dx.doi.org/10.1002/(SICI)1097-0134(199705)28:1<72::AID-PROT7>3.0.CO;2-L.

[3]   Seibert, C.M. and Raushel, F.M. (2005) Structural and Catalytic Diversity within the Amidohydrolase Superfamily. *Biochemistry* **44**:6383 – 6391.
doi: http://dx.doi.org/10.1021/bi047326v.

[4]   Aubert, S.D., Li, Y., and Raushel, F.M. (2004) Mechanism for the Hydrolysis of Organophosphates by the Bacterial Phosphotriesterase. *Biochemistry* **43**:5707 – 5715.
doi: http://dx.doi.org/10.1021/bi0497805.

[5]   Marti-Arbona, R., Xu, C., Steele, S., Weeks, A., Kuty, G.F., Seibert, C.M. and Raushel, F.M. (2006) Annotating Enzymes of Unknown Function: *N*-Formimino-L-glutamate Deiminase Is a Member of the Amidohydrolase Superfamily. *Biochemistry* **45**:1997 – 2005.
doi: http://dx.doi.org/10.1021/bi0525425.

[6]   Tabor, H. and Mehler, A.H. (1954) Isolation of *N*-Formyl-L-Glutamic Acids as an Intermediate in the Enzymatic Degradation of L-Histidine. *J. Biol. Chem.* **210**(2):559 – 568.

[7]   Nguyen, T.T., Brown, S., Fedorov, A.A., Fedorov, E.V., Babbitt, P.C., Almo, S.C. and Raushel, F.M. (2008) At the Periphery of the Amidohydrolase Superfamily: Bh0493 from *Bacillus halodurans* Catalyzes the Isomerization of D-Galacturonate to D-Tagaturonate. *Biochemistry* **47**:1194 – 1206.
doi: http://dx.doi.org/10.1021/bi7017738.

[8]   Cummings, J., Fedorov, A.A., Xu, C., Brown, S., Fedorov, E.V., Babbitt, P.C., Almo, S.C. and Raushel, F.M. (2009) Annotating Enzymes of Uncertain Function: The Deacylation of D-Amino Acids by Members of the Amidohydrolase Superfamily. *Biochemistry* **48**:6469 – 6481.
doi: http://dx.doi.org/10.1021/bi900661b.

[9]   Xiang, D.F., Patskovsky, Y., Xu, C., Meyer, A., Sauder, J.M., Burley, S.K., Almo, S.C., and Raushel, F.M. (2009) Functional Identification of Incorrectly Annotated Prolidases from the Amidohydrolase Superfamily of Enzymes. *Biochemistry* **48**:3730 – 3742.
doi: http://dx.doi.org/10.1021/bi900111q.

[10] Xiang, D.F., Xu, C., Kumaran, D., Brown, A.C., Sauder, J.M., Burley, S.K., Swami-nathan, S., Raushel, F.M. (2009) Functional Annotation of Two New Carboxypepti-dases from the Amidohydrolase Superfamily of Enzymes. *Biochemistry* **48**:4567 – 4576.
doi: http://dx.doi.org/10.1021/bi900453u.

[11] Xiang, D.F. and Raushel, F.M. (2009) unpublished observations.

[12] Hermann, J., Marti-Arbona, R., Fedorov, A.A., Fedorov, E., Almo, S.C., Shoichet, B.K. and Raushel, F.M. (2007) Structure-based activity prediction for an enzyme of unknown function. *Nature* **448**: 775 – 779.
doi: http://dx.doi.org/10.1038/nature05981.

[13] Hermann, J.C., Ghanem, E., Li, Y., Raushel, F.M., Irwin, J.J. and Shoichet, B.K. (2006) Predicting Substrates by Docking High-Energy Intermediates to Enzyme Structures. *J. Amer. Chem. Soc.* **128**:15882 – 15891.
doi: http://dx.doi.org/10.1021/ja065860f.

Understanding Enzymes as Reporters or Targets in Assays Using Quantitative High-throughput Screening (qHTS)

Douglas S. Auld[*], Natasha Thorne, Matthew B. Boxer, Noel Southall, Min Shen, Craig J. Thomas and James Inglese

NIH Chemical Genomics Center, National Institutes of Health, Bethesda, MD 20892 – 3370, U.S.A.

E-Mail: [*]dauld@mail.nih.gov

Received: 5th January 2010 / Published: 14th September 2010

Abstract

The U.S. National Institutes of Health Chemical Genomics Center (NCGC) has established a new screening paradigm, quantitative high-throughput screening (qHTS), wherein concentration-response curves (CRCs) are rapidly recorded on large compound collections (> 300,000). The data is automatically fit to the Hill equation and the CRCs are subjected to a classification scheme. This approach reduces false positive and negative rates compared to the traditional screening approaches where only a single concentration is tested and provides a pharmacological database that can be used to construct large-scale bioactivity profiles. We demonstrate how this approach was used to examine a coupled enzyme assay where the production of ATP by human pyruvate kinase M2 (PykM2) was coupled to the ATP-dependent bioluminescent enzyme, firefly luciferase (FLuc), to produce a luminescent signal. This identified chemical probes which specifically activate PykM2 while also providing a bioactivity profile of FLuc inhibitors. Examining the latter uncovered a counterintuitive phenomenon of great importance to compound discovery efforts wherein FLuc inhibitors specifically produce a non-specific luminescent response in cell-based assays.

INTRODUCTION

Screening of small molecule compound collections has been historically practiced within the pharmaceutical industry and only recently have high-throughput screening (HTS) methodologies been adopted to produce general research tools which can be applied to uncover mechanisms of biological function [1, 2]. The term "chemical biology" is now widely used in reference to identifying compounds that act as positive or negative regulators of individual gene products or signalling pathways in an effort to further the understanding of complex biological networks [3, 4]. In 2003, the U.S. National Institutes of Health (NIH) as part of the NIH Roadmap for Medical Research (http://nihroadmap.nih.gov), started the Molecular Libraries Initiative (MLI) to provide industrial-scale HTS technologies and chemical probes for basic research [5 – 7] (for more information see: mli.nih.gov). Importantly, both chemical and biological assay information is now freely accessible through the creation of the PubChem database [8] (http://pubchem.ncbi.nlm.nih.gov).

At the NIH Chemical Genomics Center (NCGC), we have endeavored to bring the expertise from both the pharmaceutical and academic communities together to enable the discovery of chemical probes and to create robust datasets that can be mined for structure-activity relationships (SAR). Compound discovery in either chemical biology or drug discovery starts with a single experiment wherein a bioassay is screened against a diverse compound collection. This experiment is critical to the entire discovery process as the results will often set the compass which guides all subsequent efforts. To provide the optimum starting point, as well as to capitalize on the most thorough use of sophisticated assays that may have taken years to develop, we have developed a novel screening paradigm, termed quantitative HTS (qHTS), where chemical libraries of about 300,000 compounds are screened at multiple concentrations so that concentration-response curves (CRCs) are measured for every compound. This provides a high quality dataset that can be used to drive chemical optimization efforts and allows construction of pharmacological databases that can be mined for bioactivity relationships. Here we present an overview of the challenges that had to be overcome to develop and implement qHTS in addition to presenting a case study of a coupled enzyme assay to highlight what has been learned while implementing this approach.

OVERVIEW OF THE CONSTRUCTION AND CLASSIFICATION OF LARGE CONCENTRATION-RESPONSE-BASED DATASETS

Physical construction of a compound dilution series can employ two general strategies: One where the compounds are titrated within the same microtiter plate to create a "horizontal" intra-plate dilution series and another where the compound dilution series is constructed in a "vertical" inter-plate manner by diluting the compound between successive microtiter plates. We selected a vertical inter-plate dilution method at the NCGC for qHTS operations [9]. In this case, plates are assayed in a manner where the first plate contains the lowest concentration of a set of compounds while subsequent plates contain the same compounds in the same well locations, but at successively increasing concentrations (Fig. 1a). This approach to large-scale concentration-series plating offers several advantages to the intra-plate method which include increased flexibility in plate usage for screening a wide variety of assay systems, and ease and speed of plate preparation [9, 10]. The vertically-developed dilution method allows one to choose between testing multiple concentrations or limiting the concentration range as necessitated by factors such as target concentration, assay sensitivity, and reagent cost (Fig. 1b).

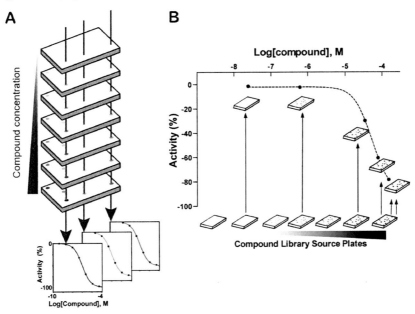

Figure 1. Construction, flexibility and customization of CRCs with qHTS. (A) Microtiter plates containing compounds in DMSO are diluted serially into DMSO to create daughter plates that contain the same compounds at the equivalent positions but at successively lower concentrations. Typically, five-fold dilutions are used for a seven point dilution series or dilutions at a fold $= \sqrt{5}$ are used to create a 14 point dilution series over the same concentration range. Shown at the bottom are CRCs obtained from the data collected from assay plates that each received 20 nL of compound solution. (B) Construction of asymmetric titration series using a set of dilution plates. Arrows represent the 20 nL transfer of compound solution from the source plates (below) to the assay plates (above). In this example, the highest concentration point is customized to contain a two-fold higher compound concentration which is achieved by a double transfer (dual arrows) from the highest concentration source plate. Adopted from Yasgar et al., 2008 [9].

Management of the 350k compound archive for qHTS at the NCGC requires processes to acquire, register, and track > 2 million sample wells, an undertaking comparable to a large pharmaceutical company [11]. The operations for this endeavor which include analytical chemistry, engineering, and software development have been described elsewhere [1, 9, 12, 13]. To minimize the cost of the compound screen and time on the robotic system, we employ miniaturized assay volumes (between 2 and 10 µL volumes) in 1,536-well microtiter plates. A typical assay protocol involves dispensing a reagent into a 1,536-well assay plate, transferring 20 nL of a DMSO compound solution to the assay plate, transfering appropriate controls to the assay plate (128 wells are reserved for controls on every microtiter plate), incubation of the assay components, and measurement of activity. Following measurement, the raw data is normalized to controls, corrected for systematic errors (due to, for example, liquid dispensing), and the quality of the data is checked by calculating the coefficient of variation (CV) between sample wells, signal to background ratio, Z-factor, and the reproducibility of potency values if control titrations were used [1, 12, 14, 15]. This provides the initial dataset used for subsequent fitting of concentration-response curves (CRCs).

Indeed one of the greatest challenges we faced in implementing the qHTS method was a means to rapidly and automatically fit CRCs to the >2 million data points generated from a single qHTS experiment. Commercial software packages lacked the ability to fit hundreds of thousands of CRCs. Even more difficult was finding a strategy to rank the CRCs in terms of curve-fit quality and compare the pharmacological responses between CRCs. To solve this, in-house software has been developed that includes identification of outlier data points, one of the largest impediments to automated curve-fitting. Commonly, outlier detection involves determining if a datapoint is significantly different (*e.g.* 3 s.d.) from the mean value. However, in concentration-response data the mean has not been determined and the mean is in fact exactly what is being fitted. Therefore, we developed a robust automated curve-fitting algorithm which results in a curve fit using a four parameter Hill equation [16, 17]:

$$y = S_0 + \frac{S_{inf} - S_0}{1 + \left(10^{\log AC_{50} - x}\right)^n}$$

Where S_0 is the activity at zero compound concentration, S_{inf} is the activity at infinite compound concentration, AC_{50} is the concentration at which the activity reaches 50% of maximal level (either activating or inhibiting). x is the tested concentration range, y yields the observed response, and n is the Hill coefficient, the slope at the AC_{50} [18]. The curve-fitting employs line interpolation based on connecting datapoints or from an expected value (*e.g.* the basal activity should be close to zero), as well as iterative curve-fitting where datapoints are masked if excluding these yields a better fit to the data. To estimate the quality of the fit to the model we use R^2, defined as

$$R^2 = 1.0 - SS_{(Hill\ fit)}/SS_{(constant\ fit)}$$

SS is the sum-of-square of the vertical distances of the points from the curve (Hill fit) or a straight line (constant fit). In this approach values of the four parameters are exhaustively sampled to find the combination having the best R^2. For this purpose, the Hill slope is constrained to values between 0.3 and 5.0, as we have found that values >5.0 have little effect on the curve fit given the number of concentration points tested in qHTS. A version of the curve-fitting software is freely available at http://www.ncgc.nih.gov/resources/software.html (and see [19] for a more detailed description).

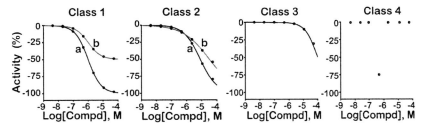

Figure 2. CRC classification scheme. Class 1 curves display two asymptotes, an inflection point, and $R^2 \geq 0.9$; subclasses 1a versus 1b are differentiated by full (> 80%) versus partial ($\leq 80\%$) response. Class 1a curves demonstrate high activity. Class 2 curves display a single left-hand asymptote and inflection point; subclasses 2a and 2b are differentiated by a maximum response and R^2 with either > 80% and > 0.9 or < 80% and < 0.9, respectively. Class 3 curves have a single left-hand asymptote, no inflection point, and a response > 3 s.d. the mean activity of the sample field; hence they are lower confidence curves. Class 4 defines those samples without any concentration–response relationship; thus they are inactive. Adopted from Shukla *et al.* 2009 [20].

The next issue to be overcome once the CRCs are obtained is a strategy to assess the quality of the curve-fit to the data as well as to compare the pharmacological parameters (*e. g.* potency and efficacy of the response) obtained between active CRCs. Common statistical tests such as the F-test have not proven useful to rank CRCs and the values of these parameters do not reflect the nature of the pharmacological response [19]. To help address these issues we developed a CRC classification scheme shown in Figure 2 [12, 20]. Classification of the CRCs in this manner provides an easily understood number reflecting both the quality of the curve-fit to the data along with information about the pharmacological response. Once the CRC classification numbers are assigned these can be used to cluster structurally related compounds to determine which of these are associated with high quality CRCs as well as those with less efficacious/potent CRCs and inactive compounds. The SAR obtained from such analysis can be immediately employed to optimize chemical probes for the intended target or to construct robust bioactivity profiles that can reveal unexpected hidden phenotypes associated with a chemical series. We illustrate both of these approaches below by examining a simple coupled enzyme assay with qHTS.

APPLICATION OF QHTS TO IDENTIFY CHEMICAL PROBES OF HUMAN PYRUVATE KINASE

Pyruvate kinase (Pyk; EC 2.7.1.40) is an allosterically regulated enzyme that functions in the final step of glycolysis to convert one molecule of ADP and phosphoenol pyruvate (PEP) to ATP and pyruvate. In cells, this reaction proceeds far from its equilibrium and PyK has one of the lowest V_{max} values among glycolytic enzymes making this one of the rate-determining steps in glycolysis. Metabolic regulatory steps catalyzed by such allosteric enzymes are responsible for the homeostasis of metabolites [21]. In cancer cells homeostasis is shifted to support rapid cellular proliferation, an abnormal state of flux for mature tissue cells [22].

Otto Warburg was first to note that cancer cells show aberrant glycolysis, producing lactate even in the presence of oxygen [23, 24]. This abnormal metabolic phenotype has become known as the "Warburg effect" (Fig. 3A) and is thought to be due at least in part to over expression of specific enzymes such as lactate dehydrogenase and the expression of a tumor specific isozyme of pyruvate kinase – PyKM2 [25 – 27].

Humans have four isozymes of PyK expressed from two genes which undergo alternative splicing [28, 29]. The Pyk L/R gene produces the liver (PyKL) and red blood cell (PyKR) forms. The pre-mRNA from the Pyk M gene is alternatively spliced in a region that encodes the allosteric activation loop of PyK (Fig. 3B) to produce the M1 isozyme (PyKM1). PyKM1 is found in most normal adult tissues and this alternative splicing results in this isoform being insensitive to regulation by fructose-1,6-bis-phosphate (FBP). In contrast, the PyKM2 isoform is activated by FBP and is only normally expressed in fetal tissue but also shows aberrant expression in tumor cells [28, 30, 31]. All tumors and cancer cell lines studied to date exhibit exclusive expression of the PyKM2 isozyme [32, 33].

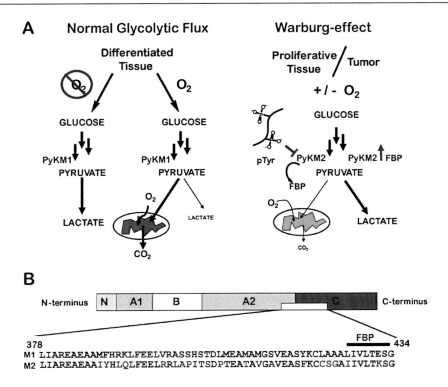

Figure 3. The Warburg effect and differential regulation of PykM1/M2. (A) Left, shows the normal state of glycolytic flux where PyKM1 is expressed in many differentiated healthy tissue types. In aerobic environments glucose is metabolized to pyruvate which is utilized by the Krebs cycle within the mitochondria; or in anaerobic environments glucose is converted to lactate. Right, illustrates the Warburg effect associated with cancer cells where glucose is converted to lactate under either aerobic or anaerobic environments. This abnormal metabolism is in part due to the expression of tumor specific enzymes such as PyKM2 which is allosterically regulated by FBP and phospho-polypeptides (pTyr) which are abundant in cancer cells from growth factor signaling pathways. The binding of pTyr is thought to down-regulate PyKM2 activity by removing the allosteric activator FBP, resulting in a build-up of glycoltyic metabolites that feed proliferation. (B) The PyKM1 and M2 isoforms arise from the M gene through alternative splicing of exons 9 and 10 located at the interface of the A and C domains. These exons encode a stretch of 56 amino acids and splicing results in amino acid differences between the two isoforms (highlighted in red for PyKM2) which includes a portion of the FBP binding pocket.

In cancer cells, the dimeric form of PyKM2 has low activity, but upon binding of FBP to its activation loop the tetrameric form of the enzyme is stabilized, leading to activation by increasing the enzyme's affinity for PEP [30, 33]. Low activity of PyKM2 in cancer cells acts to shunt key glycolytic metabolites toward biosynthetic pathways, a necessary requirement for rapidly proliferating cells such as those found in fetal tissues and tumors. Additionally, recent work has shown that replacement of PyKM2 with PyKM1 in cancer cells relieves the Warburg effect and restores a metabolic phenotype characteristic of normal cells [26]. The lung carcinoma line, H1299, when engineered to express only PyKM1 showed delayed tumor formation in nude mouse xenografts [26]. Recently, a link between growth factor signaling and PyKM2 has been shown where phosphotyrosine peptides (prevalent in cancer cells from growth factor signaling pathways) bind to PykM2 near the allosteric activation loop. Phosphotyrosine peptide binding to PykM2 results in release of FBP from the enzyme leading to further down-regulation of this critical step in glycolysis (Fig. 3A) [27]. Therefore, one approach to anti-cancer therapy would be identifying compounds that activate human PyKM2 to PyKM1 levels, which should help restore normal metabolism, shifting the glycolytic flux back toward the demand for energy rather than biosynthesis.

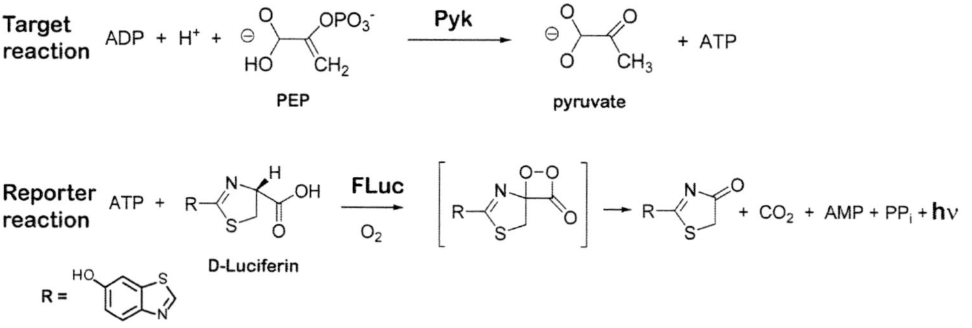

Figure 4. The pyruvate kinase-FLuc coupled assay. Three μL of substrate mix (at r.t.) in assay buffer (50 mM imidazole pH 7.2, 50 mM KCl, 7 mM MgCl$_2$, 0.01% Tween 20, 0.05% BSA) was dispensed into Kalypsys white solid bottom 1,536 well microtiter plates using a bottle-valve solenoid-based dispenser (Kalypsys). The final concentrations of substrates in the assay were 0.1 mM ADP and 0.5 mM PEP. 23 nL of compound were delivered with a 1,536-pin array tool [47] and 1 μL of enzyme mix in assay buffer (final concentration, 0.1 nM PyKM2, 50 mM imidazole pH 7.2, 0.05% BSA, 4 °C) was added. Microtiter plates were incubated at r.t. for 1 hour and 2 uL of luciferase detection mix (Kinase-Glo, Promega [48] at 4 °C protected from light) was added and luminescence was read with a ViewLux (Perkin Elmer) using a 5 second exposure/plate.

The coupled assay shown in Figure 4 was designed to find both inhibitors and activators of PyKM2 and was applied to qHTS of about 300k compounds in the Molecular Libraries Small Molecule Repository (MLSMR). The assay of ca. 2.56 million sample wells performed well with Z-factors of 0.78 ± 0.07 and CVs of $10 \pm 3\%$. This identified 777 high quality activating CRCs (0.27% of the library; Fig. 5). Inhibitory CRCs were also found (~0.02% high quality CRCs), but most of these were subsequently found to be inhibitors of the luciferase coupling enzyme (see below). The primary qHTS data is available in Pub-Chem (AIDs: 1631 and 1634).

A

Activity	Distribution	CRC Class					
		1a	1b	2a	2b	3	4
Activation	# Cmpds.	134	318	325	1541	2094	
	% library	0.05	0.11	0.11	0.55	0.74	276,309
Inhibition	# Cmpd	4	25	31	582	983	97.73%
	% library	0.00	0.01	0.01	0.21	0.35	

Figure 5. Summary of the qHTS against PyKM2.
(A) Distribution of activity by CRC class. (B) CRC's for actives from the qHTS associated with CRC classes 1a, 1b and 2a showing potency and maximum response. (C) Chemical structure of the substituted N,N'-diarylsulfonamide NCGC00030335 (**1**). (D) Selectivity assessment for **1** versus PyKM2 (open circles), PyKM1 (filled squares), PyKL (open squares), and PyKR (filled circles).

To define the SAR between compounds showing activation of PyKM2 activity we clustered structures showing high quality CRCs by structurally similarity (Fig 6). This identified a series of activators with a common substituted N,N'-diarylsulfonamide core structure. We next used this core structure to query the entire qHTS dataset to identify related structures associated with weaker CRCs or inactive compounds to allow a complete SAR from the qHTS to be developed. One of the more potent activators in this chemical series was compound **1** (Fig. 5).

After resynthesis and confirmation of chemical structure and purity, we examined the mode of action of **1** against PyKM2 through analysis of the compound's effect on the steady-state kinetics of PEP and ADP. Kinetic assays were carried out at 25 °C with 10 nM PyKM2 using [KCl] = 200 mM, $[MgCl_2]$ = 15 mM, and with either [ADP] or [PEP] = 4.0 mM through the measurement of pyruvate in a lactate dehydrogenase coupled assay [34]. As discussed, FBP is known to allosterically activate PyKM2 through induction of an enzyme state with a high affinity for PEP. Consistent with these observations, in the absence of activator, we found that PyKM2 shows low affinity for PEP ($S_{0.5} \sim 1.5$ mM). We found that in the presence of either 10 µM **1** or FBP the K_M for PEP decreased nearly 10-fold to 0.26 ± 0.08 mM or 0.1 ± 0.02 mM, respectively, with lesser effects on V_{max} (values of 245 pmols/min with or without FBP and 265 pmols/min with **1**) [34]. The addition of excess PEP abolishes the activation of PyKM2 by **1** or FBP. However, varying the concentration of ADP in the presence and absence of **1** shows that the steady-state kinetics for this substrate are not significantly affected (K_M for ADP = 0.1 mM in either condition). Akin to what has been observed for FBP, our primary lead NCGC 00030335 (**1**) lowered the K_M for PEP but had no affect on the steady-state kinetics of the ADP reaction [34].

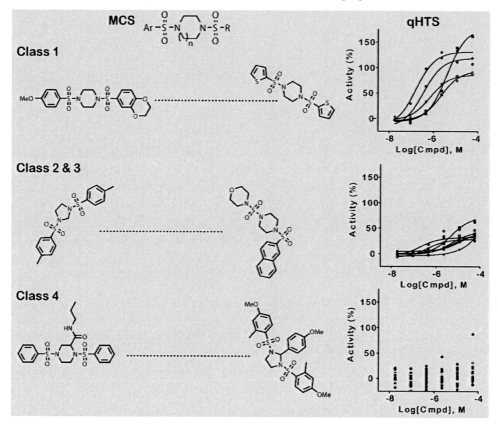

Figure 6. SAR development using primary qHTS data. In the example shown, structures associated with high quality CRCs were subjected to hierarchical clustering using Leadscope fingerprints to define a maximal common substructure (MCS). This MCS can then be used to identify similar structures associated with lower confidence CRCs or that are inactive. Example structures for the PyKM2 activator series and CRCs from the qHTS are shown.

This provided an informative dataset to initiate chemical optimization of these leads. Synthesis of nearly 300 analogs refined the SAR for this chemical series [34]. While the potency of the series was generally good (< 1 µM), the lack of aqueous solubility was a major liability for these compounds. Many analogs had solubility levels below detectable limits (e. g. **2** in Table 1). The solubility of **3** containing a seven-member ring system showed low values of 5.6 µg/mL (2.7 µM; Table 1). With an understanding of the SAR we were able to design and synthesize numerous analogues incorporating more soluble functional groups which maintained potency (**4 – 8**, Table 1; including pyridines, anilines and N-acetyl anilines). These compounds are currently being tested in cell-based assays for anti-proliferative activity and should be useful chemical probes to study the role of the Warburg effect in tumor models.

Table 1. SAR of selected N,N'-diarylsulfonamides including solubility assessment.

#	n	Ar$_1$	Ar$_2$	hPK, M2 AC$_{50}$ (μM)^a	hPK, M2 Max, Res.^{b}	Solubility^c (μM)	Solubility^c (μg/mL)
2	1	2,6-difluor-obenzene	6-(2,3-dihydroben-zo[b][1,4]dioxine)	0.065	94	<1.1	<0.5
3	2	2,6-difluor-obenzene	6-(2,3 -dihydroben-zo[b][1,4]dioxine)	0.866	120	5.6	2.7
4	1	3-aniline	2,6- difluoro-4-methoxy-benzene	0.023	87	<0.7	<0.4
5	1	3-aniline	6-(2,3 -dihydroben-zo[b][1,4]dioxine)	0.041	82	7.3	4.1
6	1	3-aniline	2,6-difluorobenzene	0.092	82	5.7	3.0
7	2	3-aniline	2,6-difluoro-4-methoxy-benzene	0.206	93	26.3	15.1
8	2	3-aniline	6-(2,3-dihydroben-zo[b][1,4]dioxine)	0.033	89	51.2	29.0

^{a}AC 50 values were determined utilizing the luminescent pyruvate kinase-luciferase coupled assay and the data represents the results from three separate experiments. ^{b}Max. Res. value represents the % activation at 57 μM of compound. ^{c}kinetic solubility analysis was performed by Analiza Inc. and are based upon quantitative nitrogen detection as described (www.analiza.com). The data represents results from three separate experiments with an average intra-assay %CV of 4.5%.

APPLICATION qHTS TO REVEAL AN UNEXPECTED CELLULAR PHENOTYPE ASSOCIATED WITH FIREFLY LUCIFERASE INHIBITORS

Enzymes are widely employed as reporters for biological assays aimed at screening large chemical libraries using HTS systems. The commercialization of firefly luciferase (FLuc) assay reagents has lead to widespread use of this enzyme. However, failure to appreciate the underlying enzymology can lead to a misuse of this powerful assay technology resulting in misinterpreted results. In drug discovery the HTS assay often sets the course of action for a development program that can take 10 to 15 years to transpire at a cost of US$ 800 M. Initiating this process with well-considered leads may significantly improve the efficiency of drug discovery. Therefore, at the NCGC we have endeavored to understand the mechanisms underlying assay interference, a major impediment to the identification of quality leads. The FLuc enzyme is one of the most common reporter enzymes used in bioassays [35] and we sought to determine how this enzyme's activity is modulated by small molecules found in typical HTS compound collections.

Using qHTS we determined the concentration-response behavior for > 70,000 compounds in the MLSMR against the ATP-dependent luciferase from the firefly *Photinus pyralis* (FLuc; PubChem AID: 411; EC 1.13.12.7). Approximately 3% of the library showed inhibitory activity while none of the compounds caused activation of enzyme activity. Given that the activity relevant to the target being explored in a typical HTS is ~0.01 – 0.1%, this level of inhibition is highly significant and will easily confound the analysis of actives from a screen if steps are not taken to address this "off-target" activity.

Through examination of this profile we were able to define the SAR for prominent luciferase inhibitor series [35]. We identified several structural classes of inhibitors which included compounds showing structural similarity to the benzthiazole core of the substrate D-luciferin. Scaffolds likely to mimic the D-luciferin substrate were often found to be competitive with D-luciferin, but not ATP (Fig. 7a), while others such as those containing a quinoline core were found to be ATP competitive as well (Fig. 7b). Compounds showing no structural similarity to either ATP or D-luciferin but possessing potent inhibitory activity (such as a series of 3,5-diaryl oxadiazoles) were also identified (Fig. 7c).

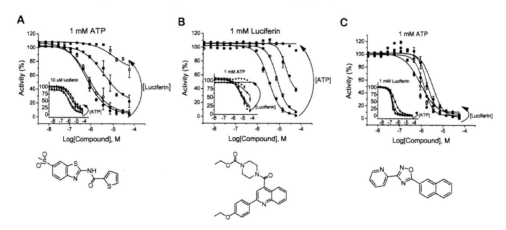

Figure 7. Representative luciferase inhibtors from the FLuc qHTS. Shown is the dependence of inhibition on substrate concentration for three representative compounds. Luciferin or ATP was varied at (●) 1 μM, (O) 10 μM, (■) 100 μM, and (□) 1000 μM. (A) Substrate variation data for a benzthiazole representing a compound likely to mimic the D-luciferin substrate. The graph shows that inhibition can be relieved upon increasing the luciferin concentration, and the inset graph shows that ATP variation has little effect on the inhibition (holding the luciferin concentration constant at 10 μM, approximately the K_M). (B) Substrate variation data for a representative quinoline. The graph shows that inhibition can be relieved when either the ATP or the luciferin concentration (inset graph) is increased. (C) Substrate variation data for a 3,5-diaryl oxadiazole. The graph shows that the inhibition remains relatively constant when varying either the luciferin or the ATP concentration (inset graph). Adapted from [35].

Our profile used a cell-free system with purified FLuc to define inhibitors. However, FLuc is commonly used in cell-based reporter-gene assays because the luminescent response provides a sensitive assay signal. Due to the relatively short protein half-life of FLuc this reporter can be used to measure dynamic responses in gene regulation where either increases or decreases in luciferase activity are measured [36]. Enzymes can be stabilized by inhibitors [37] when an E•I complex is more resistant to degradation than the free enzyme. For this to occur, the reporter enzyme (E) expressed in a cell must bind to a cell-penetrating inhibitor (I) leading to an E•I complex within the cells that stabilizes the enzyme. This may occur, for example, due to a conformational change upon inhibitor binding that stabilizes the enzyme from degradation. Therefore, assuming no effect due to transcription or translation, we would expect the total amount of active enzyme (E) at any time to depend on the concentration of E•I, which, itself is dependent on the affinity of the complex within the cells. The degradation of the free enzyme (E_f) and E•I can be shown as:

$$E* \xleftarrow{k_2} E \cdot I \xleftrightarrow{K_I} + E_f \xrightarrow{k_1} E*$$

Where E* is degraded inactive enzyme and the rate of degradation is

$$-\frac{dE}{dt} = k_1 \left[E_f\right] + k_2 [E \cdot I](k2 < k1)$$

The amount of intact active enzyme (E) produced will be modulated by term ($1+I/K_I$). When the ratio of I to K_I is high, the slower degrading E•I complex is maximized so that degradation of the enzyme is minimized. This simple model predicts that at different concentrations of I the amount of active enzyme available will follow a CRC with an apparent EC_{50} that parallels the affinity of inhibitor for the enzyme (K_I; Fig. 8A). However, the final apparent pharmacology that is observed in the assay will depend on the sum of two opposing inhibitor responses: the positive increase in enzyme levels due to the stabilization by the inhibitor in the cells and the opposing inhibitory effect on the enzyme reaction during the detection step where excess substrates are added. Depending on the mode of action of the inhibitor (e.g. reversible or irreversible, competitive or noncompetitive) and the configuration of the assay different cases are possible and these are illustrated in Figure 8 [38]. Inhibitors that are not completely removed by the detection reagents can show bell-shaped CRCs, a phenomenon that is typically attributed to cytotoxicity in cell-based assays although reporter inhibition can lead to the same response.

We have experimentally demonstrated that inhibitor-based stabilization of FLuc occurs for representative chemotypes from FLuc our inhibitor profiling efforts [35, 39]. We measured the CRCs for luciferase activity before and after treating cells with cycloheximide for 24 hours. The same compounds were also measured in an assay using purified FLuc and K_M levels of substrates to confirm the inhibitory effect of these compounds. In these experiments, we observed apparent activation of the luciferase signal upon addition of a reporter-gene detection cocktail containing excess luciferase substrates which we found can occur at concentration ranges typically used in HTS ($1-10\,\mu M$; Fig. 8C). Examination of the stability of the signal after cycloheximide treatment showed a slower rate of activity decay from samples treated with inhibitor compared to samples without inhibitor [39]. Plots of the relative amount of luciferase activity remaining after a 24 h treatment with cycloheximide

(Fig. 8C, red line) showed a CRC that mirrored the inhibition against the purified enzyme. These parallel but opposite responses are expected based on stabilization of an E•I complex, where the potency for activation response within cells mirrors the inhibition potency against the purified enzyme (as described above) and strongly supports that increased FLuc activity is due to inhibitor based-stabilization of the FLuc enzyme.

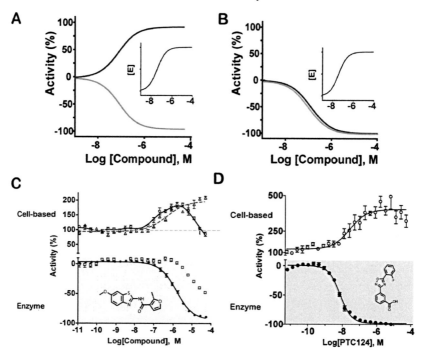

Figure 8. Post-translational inhibitor-based stabilization of FLuc in cell-based assays. Shown is what should be theoretically observed for reporter activity in the cell-based assays (blue lines) and the activity against the reporter enzyme determined in a biochemical assay where substrates are at low concentrations (e.g.,~ K_M; orange lines). Insets illustrate the increase in intracellular enzyme concentration in the assay as a result of inhibitor-based stabilization. (A) If the cells are treated with a reversible competitive inhibitor of luciferase, after interaction and stabilization of the luciferase protein (inset), the compound can be competed off the luciferase enzyme after cell lysis by the addition of a detection reagent containing excess substrates. In this case, apparent activation is observed in the cell-based assay (blue curve) that mirrors the inhibition of the reporter enzyme in a cell-free, biochemical assay. Alternatively, activity from reversible noncompetitive inhibitors may be detected by removal of the inhibitor through cell wash steps prior to detection. (B) If the cells are treated with a compound that acts as an irreversible or a noncompetitive inhibitor of the luciferase enzyme (orange curve) where the inhibitor is not removed before detection, stabilization of the reporter enzyme still occurs (inset), but in this scenario inhibition is likely observed (blue curve) because the inhibitor is not displaced from the enzyme during detection. Adopted from ref [38]. (C) More complex apparent activation behavior such as bell-shaped concentration–response curves may also occur as illustrated

here for luciferase activity measured from HEK293 cells expressing FLuc and treated with compound for 24 h (top graph) (O, black fitted line) or the remaining luciferase activity following 24-h treatment with cycloheximide (red fitted line). Bottom graphs depict the cell-free luciferase activity determined using purified luciferase assayed with a commercial reporter gene detection cocktail (SteadyGlo^{TM}, Promega □) or using K_M concentrations of luciferase substrates (●). See [39]. (D) (Upper) FLuc activity from Grip-Tite 293 cells transfected with the pFLuc190^{UGA} construct and treated with PTC 124 for 24 h (PTC 124; open circles) shows concentration-dependent activation. (Lower) Activity of purified FLuc enzyme (with K_M concentrations of substrates) shows concentration-dependent inhibition with compound treatment (filled circles). In this case, the inhibitor was removed from the cells by wash steps prior to detection (see [38] for further details).

To examine the prevalence of this phenomenon among HTS assays using FLuc, we examined the enrichment of luciferase inhibitors in 100 FLuc assays found in PubChem (Fig. 9). As our luciferase qHTS identified a frequency of luciferase inhibitors of 3%, active sets within PubChem containing only 3% luciferase inhibitors would not be considered enriched in luciferase inhibitors. However, an HTS active set found to contain, for example, 30% luciferase inhibitors is enriched 10-fold. As anticipated, we found an enrichment of luciferase inhibitors in luciferase-based assays designed to identify compounds acting as inhibitors in either biochemical or cell-based assay formats (yellow and blue bars in Figure 9, respectively). However, reporter-gene assays targeting an increase in luciferase activity also displayed a similar enrichment of luciferase inhibitors within active data sets (Fig. 9, orange bar). In contrast, no enrichment in luciferase inhibitors was observed for reporter-gene assays employing unrelated enzymes such as β-lactamase (EC 3.5.2.6), non-enzymatic reporters such as GFP, or in other assay formats using fluorescence or chemiluminescent detection strategies (Fig. 9, light grey bar). Therefore, the prevalence of luciferase inhibitors within PubChem (which represents a typical diverse screening collection) and their enrichment in luciferase reporter-gene assays aimed at activation, suggests that inhibitor-mediated reporter stabilization is a common artifact associated with cell-based assay formats using FLuc.

One of the most potent and prominent chemical series of FLuc inhibitors was contained within a subset of 3,5-diaryl-oxadiazoles. A 3,5-diaryl-oxadiazole that has received significant attention due to its purported activity as a nonsense codon suppressor and is in clinical trials for DMD and CF is PTC 124 (Fig. 8D, [40 – 42]. PTC 124 was identified as a compound that apparently increased the frequency of read-through of a premature nonsense mutation in a FLuc gene [42]. The FLuc cell-based assay used to test compounds for this activity relied on the detection of increased luciferase activity, presumably due to increased translation of full-length FLuc protein as a result of nonsense codon suppression activity by the small molecule. In the initial characterization of PTC 124, FLuc activity was shown to increase while FLuc mRNA levels were unchanged in the presence of PTC 124 [42]. Though consistent with this data, the possibility of post-translational reporter stabilization was not investigated. Therefore, we examined if the activity of PTC 124 and other 3,5-diaryl-oxadiazole analogs was attributed to post-translational stabilization of the FLuc reporter itself [38]. We found that PTC 124 and related analogs, as well as structurally unrelated FLuc inhibitors, all showed similar levels of activation in a FLuc-dependent reporter assay of nonsense codon suppression where the apparent activation paralleled inhibition against the purified FLuc enzyme, consistent with formation of a stable cellular E•I complex (for example CRCs for PTC 124 see Figure 8D, and see [38]). Further, we found no apparent nonsense codon suppression activity for PTC 124 when employing an orthogonal luminescent reporter enzyme *Renilla reniformis* luciferase (RLuc; EC 1.13.12.5) in an analogous assay. RLuc enzymatic activity was refractory to inhibition by PTC 124 and structurally related analogs [38]. Based on these lines of evidence we conclude that the initial discovery of PTC 124 was biased by post-translational inhibitor-based reporter stabilization.

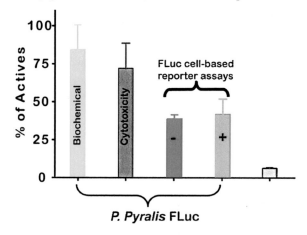

Figure 9. Percentage of luciferase inhibitors within hits from 100 PubChem assays. The PubChem active list from each assay was compared to the luciferase qHTS activity and all compounds showing inhibitory CRCs in the luciferase assay were used to calculate the percentage. Bars show the average and s.d. for the percentage of FLuc inhibitors found in each assay type. Data for figure is taken from from ref [39]. Yellow bar, *P. pyralis* luciferase-based biochemical assays (n = 6); Grey bar, cytotoxicity assays using *P. pyralis* (Perkin Elmer detection reagent; n = 4); blue bar, cell-based reporter-gene assays scored for inhibition (n = 9); Orange bar, cell-based reporter-gene assays scored for activation (n = 11); light grey bar other assay types including cell-based reporter-gene assays using β-lactamase, FRET-based assays, absorbance-based assays, and fluorescent-based assays (n = 70).

Our study of luciferase inhibitors has provided essential guidance for application of this enzyme in assay development and HTS [43]. In general, whether or not apparent activation is detected will depend on how much E•I is formed within the cells that results in stabilization and how efficiently the inhibitor is competed off in the presence of detection reagents [38, 44, 45]. Given these factors, this effect will be most commonly observed in assays which show an initially low basal level of luciferase expression and where a competitive inhibitor functions to stabilize the reporter. Use of commercial detection reagents containing excess FLuc substrates/co-factors readily removes the competitive inhibitor to allow detection of FLuc activity which is easily misinterpreted as apparent activation of the reporter gene.

We recommend a few counter-screens to detect if inhibitor-based reporter stabilization is operating in FLuc cell-based assays. One such counter-screen is a purified FLuc enzyme assay using defined (K_M levels) of substrates to properly measure inhibition. An attenuated luciferase expression system, independent of the biology being targeted, has been employed as a counter-screen [43]. In addition, a paired cell-based assay that is identical to the FLuc assay except that an unrelated reporter such as RLuc, GFP, or β-lactamase is substituted can be used. The use of such orthogonal assays [46], which are based on different reporters expressed in a common cell type, provide a method to detect if the compounds are showing genuine target-mediated activity.

CONCLUSION

Exploration of the PyKM2-FLuc coupled enzyme assay has lead to two important results. One is a chemical probe, a specific activator of the tumor specific isoform PyKM2, which will be useful in studying the Warburg effect in cancer and also serves as lead for anti-cancer drug development. The second is a compound profile which allowed us to connect luciferase inhibition to an unexpected phenotype where reporter inhibitors lead to apparent gene activation responses in FLuc cell-based assays. The counterintuitive finding that inhibitors of reporters can appear as activators in cell-based assays has not been widely appreciated although this type of 'off-target' response can bias the SAR of early phase compounds and misdirect subsequent efforts. These findings demonstrate the importance of basic research into HTS assay design and interpretation in which the basic principles of enzymology and pharmacology must be applied.

REFERENCES

[1] Inglese, J., Auld, D.S. (2009). High Throughput Screening (HTS) Techniques: Overview of Applications in Chemical Biology. *In*: Wiley Encyclopedia of Chemical Biology, Volume 2, T. Begley, ed. (Hoboken: John Wiley & Sons, Inc.), pp. 260 – 274.

[2] Stockwell, B.R. (2004). Exploring biology with small organic molecules. *Nature* **432**: 846 – 854.
doi: http://dx.doi.org/10.1038/nature03196.

[3] Collins, F.S., Green, E.D., Guttmacher, A.E., and Guyer, M.S. (2003). A vision for the future of genomics research. *Nature* **422**:835 – 847.
doi: http://dx.doi.org/10.1038/nature01626.

[4] Spring, D.R. (2005). Chemical genetics to chemical genomics: small molecules offer big insights. *Chem. Soc. Rev.* **34**: 472 – 482.
doi: http://dx.doi.org/10.1039/b312875j.

[5] Austin, C.P., Brady, L.S., Insel, T.R., and Collins, F.S. (2004). NIH Molecular Libraries Initiative. *Science* **306**:1138 – 1139.
doi: http://dx.doi.org/10.1126/science.1105511.

[6] Lazo, J.S. (2006). Roadmap or roadkill: a pharmacologist's analysis of the NIH Molecular Libraries Initiative. *Mol. Interv.* **6**:240 – 243.
doi: http://dx.doi.org/10.1124/mi.6.5.1.

[7] Lazo, J.S., Brady, L.S., and Dingledine, R. (2007). Building a pharmacological lexicon: small molecule discovery in academia. *Mol. Pharmacol.* **72**:1 – 7.
doi: http://dx.doi.org/10.1124/mol.107.035113.

[8] Richard, A.M., Gold, L.S., and Nicklaus, M.C. (2006). Chemical structure indexing of toxicity data on the internet: moving toward a flat world. *Curr. Opin. Drug. Discov. Devel.* **9**:314 – 325.

[9] Yasgar, A., Shinn, P., Jadhav, A., Auld, D., Michael, S., Zheng, W., Austin, C.P., Inglese, J., and Simeonov, A. (2008). Compound Management for Quantitative High-Throughput Screening. *JALA Charlottesv Va* **13**:79 – 89.

[10] Auld, D., Inglese, J., Jadhav, A., Austin, C.P., Sittampalam, G.S., Montrose-Rafizadeh, C., McGee, J.E., and Iversen, P.W. (2007). HTS technologies to facilitate chemical genomics. *European Pharmaceutical Review* **12**:53 – 63.

[11] Archer, J.R. (2004). History, evolution, and trends in compound management for high throughput screening. *Assay Drug Dev. Technol.* **2**:675 – 681.
doi: http://dx.doi.org/10.1089/adt.2004.2.675.

[12] Inglese, J., Auld, D.S., Jadhav, A., Johnson, R.L., Simeonov, A., Yasgar, A., Zheng, W., and Austin, C.P. (2006). Quantitative high-throughput screening: A titration-based approach that efficiently identifies biological activities in large chemical libraries. *Proc. Natl. Acad. Sci. U.S.A.* **103**:11473 – 11478.
doi: http://dx.doi.org/10.1073/pnas.0604348103.

[13] Michael, S., Auld, D., Klumpp, C., Jadhav, A., Zheng, W., Thorne, N., Austin, C.P., Inglese, J., and Simeonov, A. (2008). A robotic platform for quantitative high-throughput screening. *Assay Drug Dev. Technol.* **6**:637 – 657.
doi: http://dx.doi.org/10.1089/adt.2008.150.

[14] Eastwood, B.J., Farmen, M.W., Iversen, P.W., Craft, T.J., Smallwood, J.K., Garbison, K.E., Delapp, N.W., and Smith, G.F. (2006). The minimum significant ratio: a statistical parameter to characterize the reproducibility of potency estimates from concentration-response assays and estimation by replicate-experiment studies. *J. Biomol. Screen.* **11**:253 – 261.
doi: http://dx.doi.org/10.1177/1087057105285611.

[15] Iversen, P.W., Eastwood, B.J., Sittampalam, G.S., and Cox, K.L. (2006). A comparison of assay performance measures in screening assays: signal window, Z' factor, and assay variability ratio. *J. Biomol. Screen.* **11**:247 – 252.
doi: http://dx.doi.org/10.1177/1087057105285610

[16] Hill, A.V. (1910). The possible effects of the aggregation of the molecules of haemoglobin on its dissociation curves. *Journal of Physiology* **40**:4 – 7.

[17] Motulsky, H., Christopoulos, A. (2004). Fitting models to biological data using linear and nonlinear regression: a practical guide to curve fitting (New York: Oxford University Press).

[18] Weiss, J.N. (1997). The Hill equation revisited: uses and misuses. *FASEB J.* **11**:835 – 841.

[19] Southall, N.T., Jadhav, A., Huang, R., Nguyen, T., and Wang, Y. (2009). Enabling the Large Scale Analysis of Quantitative High Throughput Screening Data. In Handbook of Drug Screening. Second Edition, Seethala, R., Zhang, L. Eds. (New York: Taylor and Francis), pp. 442 – 462.

[20] Shukla, S.J., Nguyen, D.T., Macarthur, R., Simeonov, A., Frazee, W.J., Hallis, T.M., Marks, B.D., Singh, U., Eliason, H.C., Printen, J., Austin, C.P., Inglese, J., and Auld, D.S. (2009). Identification of pregnane X receptor ligands using time-resolved fluorescence resonance energy transfer and quantitative high-throughput screening. *Assay Drug Dev. Technol.* **7**:143 – 169.
doi: http://dx.doi.org/10.1089/adt.2009.193.

[21]   Hofmeyr, J.S., and Cornish-Bowden, A. (2000). Regulating the cellular economy of supply and demand. *FEBS Lett.* **476**:47 – 51.
doi: http://dx.doi.org/10.1016/S0014-5793(00)01668-9.

[22]   Feron, O. (2009). Pyruvate into lactate and back: from the Warburg effect to symbiotic energy fuel exchange in cancer cells. *Radiother. Oncol.* **92**:329 – 333.
doi: http://dx.doi.org/10.1016/j.radonc.2009.06.025.

[23]   Warburg, O. (1956). On respiratory impairment in cancer cells. *Science* **124**:269 – 270.

[24]   Warburg, O. (1956). On the origin of cancer cells. *Science* **123**:309 – 314.
doi: http://dx.doi.org/10.1126/science.123.3191.309.

[25]   Van der Heiden, M.G., Cantley, L.C., and Thompson, C.B. (2009). Understanding the Warburg effect: the metabolic requirements of cell proliferation. *Science* **324**:1029 – 1033.
doi: http://dx.doi.org/10.1126/science.1160809.

[26]   Christofk, H.R., Van der Heiden, M.G., Harris, M.H., Ramanathan, A., Gerszten, R.E., Wei, R., Fleming, M.D., Schreiber, S.L., and Cantley, L.C. (2008). The M2 splice isoform of pyruvate kinase is important for cancer metabolism and tumour growth. *Nature* **452**:230 – 233.
doi: http://dx.doi.org/10.1038/nature06734.

[27]   Christofk, H.R., Van der Heiden, M.G., Wu, N., Asara, J.M., and Cantley, L.C. (2008). Pyruvate kinase M2 is a phosphotyrosine-binding protein. *Nature* **452**:181 – 186.

[28]   Noguchi, T., Inoue, H., and Tanaka, T. (1986). The M1- and M2-type isozymes of rat pyruvate kinase are produced from the same gene by alternative RNA splicing. *J. Biol. Chem.* **261**:13807 – 13812.
doi: http://dx.doi.org/10.1038/nature06667.

[29]   Noguchi, T., Yamada, K., Inoue, H., Matsuda, T., and Tanaka, T. (1987). The L- and R-type isozymes of rat pyruvate kinase are produced from a single gene by use of different promoters. *J. Biol. Chem.* **262**:14366 – 14371.

[30]   Dombrauckas, J.D., Santarsiero, B.D., and Mesecar, A.D. (2005). Structural basis for tumor pyruvate kinase M2 allosteric regulation and catalysis. *Biochemistry* **44**:9417 – 9429.
doi: http://dx.doi.org/10.1021/bi0474923.

[31]   Yamada, K., and Noguchi, T. (1999). Regulation of pyruvate kinase M gene expression. *Biochem. Biophys. Res. Commun.* **256**:257 – 262.
doi: http://dx.doi.org/10.1006/bbrc.1999.0228.

[32] DeBerardinis, R.J., Lum, J.J., Hatzivassiliou, G., and Thompson, C.B. (2008). The biology of cancer: metabolic reprogramming fuels cell growth and proliferation. *Cell Metab.* **7**:11 – 20.
doi: http://dx.doi.org/10.1016/j.cmet.2007.10.002.

[33] Deberardinis, R.J., Sayed, N., Ditsworth, D., and Thompson, C.B. (2008). Brick by brick: metabolism and tumor cell growth. *Curr. Opin. Genet. Dev.* **18**:54 – 61.
doi: http://dx.doi.org/10.1016/j.gde.2008.02.003.

[34] Boxer, M.B., Jiang, J.-K., Van der Heiden, M.G., Shen, M., Skoumbourdis, A.P., Southall, N., Veith, H., Leister, W., Austin, C.P., Park, H.W., Inglese, J., Cantley, L.C., Auld, D.S., Thomas, C.J. (2009) Activators of the Tumor Cell Specific M2 Isoform of Pyruvate Kinase. Part 1: Substituted *N,N'*-Diarylsulfonamides. *J. Med. Chem.* **53**:1048 – 1055.
doi: http://dx.doi.org/10.1021/jm901577g.

[35] Auld, D.S., Southall, N.T., Jadhav, A., Johnson, R.L., Diller, D.J., Simeonov, A., Austin, C.P., and Inglese, J. (2008). Characterization of chemical libraries for luciferase inhibitory activity. *J. Med. Chem.* **51**:2372 – 2386.
doi: http://dx.doi.org/10.1021/jm701302v.

[36] Fan, F., and Wood, K.V. (2007). Bioluminescent assays for high-throughput screening. *Assay Drug Dev. Technol.* **5**:127 – 136.
doi: http://dx.doi.org/10.1089/adt.2006.053.

[37] Thompson, P.A., Wang, S., Howett, L.J., Wang, M.M., Patel, R., Averill, A., Showalter, R.E., Li, B., and Appleman, J.R. (2008). Identification of ligand binding by protein stabilization: comparison of ATLAS with biophysical and enzymatic methods. *Assay Drug Dev. Technol.* **6**:69 – 81.
doi: http://dx.doi.org/10.1089/adt.2007.100.

[38] Auld, D.S., Thorne, N., Maguire, W.F., and Inglese, J. (2009). Mechanism of PTC 124 activity in cell-based luciferase assays of nonsense codon suppression. *Proc. Natl. Acad. Sci. U.S.A.* **106**:3585 – 3590.
doi: http://dx.doi.org/10.1073/pnas.0813345106.

[39] Auld, D.S., Thorne, N., Nguyen, D.T., and Inglese, J. (2008). A specific mechanism for nonspecific activation in reporter-gene assays. *ACS Chem. Biol.* **3**:463 – 470.
doi: http://dx.doi.org/10.1021/cb8000793.

[40] Du, M., Liu, X., Welch, E.M., Hirawat, S., Peltz, S.W., and Bedwell, D.M. (2008). PTC 124 is an orally bioavailable compound that promotes suppression of the human CFTR-G542X nonsense allele in a CF mouse model. *Proc. Natl. Acad. Sci. U.S.A.* **105**:2064 – 2069.
doi: http://dx.doi.org/10.1073/pnas.0711795105.

[41]   Kerem, E., Hirawat, S., Armoni, S., Yaakov, Y., Shoseyov, D., Cohen, M., Nissim-Rafinia, M., Blau, H., Rivlin, J., Aviram, M., Elfring, G.L., Northcutt, V.J., Miller, L.L., Kerem, B., and Wilschanski, M. (2008). Effectiveness of PTC 124 treatment of cystic fibrosis caused by nonsense mutations: a prospective phase II trial. *Lancet* **372**:719 – 727.
doi: http://dx.doi.org/10.1016/S0140-6736(08)61168-X.

[42]   Welch, E.M., Barton, E.R., Zhuo, J., Tomizawa, Y., Friesen, W.J., Trifillis, P., Paushkin, S., Patel, M., Trotta, C.R., Hwang, S., Wilde, R.G., Karp, G., Takasugi, J., Chen, G., Jones, S., Ren, H., Moon, Y.C., Corson, D., Turpoff, A.A., Campbell, J.A., Conn, M.M., Khan, A., Almstead, N.G., Hedrick, J., Mollin, A., Risher, N., Weetall, M., Yeh, S., Branstrom, A.A., Colacino, J.M., Babiak, J., Ju, W.D., Hirawat, S., Northcutt, V.J., Miller, L.L., Spatrick, P., He, F., Kawana, M., Feng, H., Jacobson, A., Peltz, S.W., and Sweeney, H.L. (2007). PTC 124 targets genetic disorders caused by nonsense mutations. *Nature* **447**:87 – 91.
doi: http://dx.doi.org/10.1038/nature05756.

[43]   Lyssiotis, C.A., Foreman, R.K., Staerk, J., Garcia, M., Mathur, D., Markoulaki, S., Hanna, J., Lairson, L.L., Charette, B.D., Bouchez, L.C., Bollong, M., Kunick, C., Brinker, A., Cho, C.Y., Schultz, P.G., and Jaenisch, R. (2009). Reprogramming of murine fibroblasts to induced pluripotent stem cells with chemical complementation of Klf4. *Proc. Natl. Acad. Sci. U.S.A.* **106**:8912 – 8917.
doi: http://dx.doi.org/10.1073/pnas. 0903860106.

[44]   Inglese, J., Thorne, N., and Auld, D.S. (2009). Reply to Peltz et al: Post-translational stabilization of the firefly luciferase reporter by PTC 124 (Ataluren). *Proc. Natl. Acad. Sci. U.S.A.* **106**:E65.
doi: http://dx.doi.org/10.1073/pnas.0905457106.

[45]   Peltz, S.W., Welch, E.M., Jacobson, A., Trotta, C.R., Naryshkin, N., Sweeney, H.L., and Bedwel, D.M. (2009). Nonsense suppression activity of PTC 124 (ataluren). *Proc. Natl. Acad. Sci. U.S.A.* **106**:E64.
doi: http://dx.doi.org/10.1073/pnas.0901936106.

[46]   Inglese, J., Johnson, R.L., Simeonov, A., Xia, M., Zheng, W., Austin, C.P., and Auld, D.S. (2007). High-throughput screening assays for the identification of chemical probes. *Nat. Chem. Biol.* **3**:466.
doi: http://dx.doi.org/10.1038/nchembio.2007.17.

[47]   Cleveland, P.H., and Koutz, P.J. (2005). Nanoliter dispensing for uHTS using pin tools. *Assay Drug Dev. Technol.* **3**:213 – 225.
doi: http://dx.doi.org/10.1089/adt.2005.3.213.

[48] Auld, D.S., Zhang, Y.Q., Southall, N.T., Rai, G., Landsman, M., Maclure, J., Lange-vin, D., Thomas, C.J., Austin, C.P., and Inglese, J. (2009). A Basis for Reduced Chemical Library Inhibition of Firefly Luciferase Obtained from Directed Evolution. *J. Med. Chem.* **52**:1450 – 1458.
doi: http://dx.doi.org/10.1021/jm8014525.

How Streptococci Make Isoprenoids

Scott T. Lefurgy* and Thomas S. Leyh

Department of Microbiology & Immunology,
The Albert Einstein College of Medicine,
1300 Morris Park Ave., Bronx, NY 10461, U.S.A.

E-Mail: *scott.lefurgy@einstein.yu.edu

Received: 16th June 2010 / Published: 14th September 2010

Abstract

Isoprenoids are the set of ~ 25,000 unique compounds based on the
ubiquitous C_5 donor isopentenyl diphosphate (IPP), including qui-
nones, steroid hormones, bile acids, protein membrane anchors and
secondary metabolites. Streptococci and other gram-positive bacteria
produce IPP *via* the mevalonate pathway, whose function is required
for the respiratory pathogen *Streptococcus pneumoniae* to survive in
lung and serum. With the discovery of potent selective feedback inhibition by the metabolite diphosphomevalonate (DPM), our laboratory
has positioned the pneumococcal mevalonate pathway as a novel target
for clinical intervention against an organism that claims the lives of
over 4000 people daily. Our studies have revealed unique features of
each of each of the three GHMP family kinases that comprise the
pathway, including potent allosteric inhibition, a catalytic switch, and
a concerted elimination mechanism-informing the design of antibiotics
that can simultaneously inhibit multiple steps in a single pathway.

Introduction

Streptococcus pneumoniae kills over 1 million people each year worldwide, mostly children
and the elderly, and is the primary bacterial cause of pneumonia, meningitis and otitis media
[1, 2]. Antibiotic resistance remains a major problem in treating infections, and multiple-
drug resistance rates as high as 95% are seen in some countries [3]. Despite the successful
introduction in 2000 of a vaccine covering seven of the most prevalent and infectious of the

> 100 subtypes of pneumcoccus [5], nonvaccinated subtypes are rapidly filling the biological niche created by the vaccine and becoming more virulent [6]. There is an unequivocal need for new strategies to fight this pernicious bacterium.

Isoprenoid biosynthesis has recently emerged as a new target for antibiotic development. The isoprenoids are a class of ~ 25,000 unique compounds composed of a single building-block, isopentenyl diphosphate (IPP). The C_5 isoprene units of IPP are concatenated and then converted either to universal essential cofactors, vitamins, steroids and a host of secondary metabolites, or attached to other biomolecules, such tRNAs or proteins-facilitating association of the latter with membranes. Most eubacteria synthesize IPP starting from glyceraldehyde-3-phosphate and pyruvate using the methylerythritol phosphate pathway [7]. Gram-positive bacteria, including *S. pneumoniae*, archaebacteria and eukaryotes make IPP starting from acetyl-CoA *via* the mevalonate pathway, which converts mevalonate to IPP in three steps catalyzed by GHMP family kinases: mevalonate kinase (MK), phosphomevalonate kinase (PMK) and diphosphomevalonate decarboxylase (DPM-DC) (Fig. 1). A functional mevalonate pathway is essential for survival of *S. pneumoniae* in lung and serum [8].

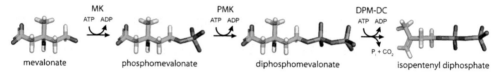

Figure 1. *The Mevalonate Pathway.* Mevalonate kinase (MK), phosphomevalonate kinase (PMK) and diphospho-mevalonate decarboxylase (DPM-DC).

MEVALONATE KINASE

Feedback inhibition is a hallmark of metabolic pathways. Human mevalonate kinase is potently inhibited by end-products farnesyl diphosphate (C_{15}) and geranylgeranyl diphosphate (C_{20}) [9, 10]. We were therefore surprised to discover that the the MK homologue in *S. pneumoniae* is subject to inhibition, not by end-products, but by the intermediate DPM. The MK and PMK reactions each produce ADP, which can be stoichiometrically coupled to the oxidation of NADH by a pyruvate kinase/lactate dehydrogenase coupling system-thus providing a continuous optical assay for enzyme activity (Fig. 2). When MK and PMK are added in successive steps to an assay mixture containing mevalonate, one equivalent of DPM is produced. However, when MK and PMK are added simultaneously, very little product is formed. This result suggested that the DPM produced during the initial stages of the reaction was inhibiting the first enzyme, MK. Incubation of MK in the presence of DPM strongly inhibited turnover, confirming this hypothesis. Titration of DPM against MK showed that its potency was quite strong (IC_{50} ~ 400 nM). Tests of human MK yielded no DPM inhibition at levels up to 25 µM, confirming that *S. pneumoniae* MK could be specifically targeted [11].

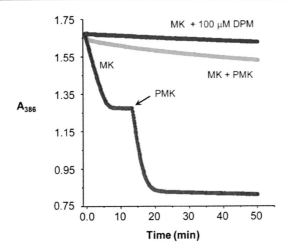

Figure 2. *Discovery of MK Inhibition by DPM*. Reaction progress curves for reactions in which enzymes were added sequentially or simultaneously. ADP produced by MK or PMK is stoichiometrically coupled to oxidation of NADH by a pyruvate kinase/ lactate dehydrogenase coupling system, resulting in a decrease in absorbance at 386 nm.

An important consideration when developing an antibiotic that targets a metabolic pathway is the mechanism of inhibition. An allosteric mechanism of pathway inhibition – in which the inhibitory ligand binds distal to active site – has an advantage over simple occlusion of the active site, in that inhibition cannot be overcome by the thermodynamic push that accompanies a buildup of the metabolite (i. e., the substrate) just upstream of the inhibited step. MK inhibition by DPM was investigated using initial-rate experiments that simulate this condition and can thus distinguish between mechanisms: the maximal reaction velocity at (theoretically) infinite substrate concentration was determined as a function of inhibitor concentration. If DPM binds to the active site, its inhibitory effects (at any concentration of inhibitor) will be completely irrelevant at infinite substrate concentration and have no impact on the maximal velocity. We observed that DPM reduced this maximal velocity – with equal potency *versus* both mevalonate and ATP – indicating that DPM must bind to an allosteric site. This result was confirmed by the observation that DPM had an identical affinity for MK both in the presence and absence of saturating concentrations of mevalonate and AMPPNP (a non-hydrolyzable ATP analogue) in an equilibrium binding study. If DPM bound at the active site, its apparent affinity should have been altered by the presence of competing ligands. The stoichiometry of DPM binding to the MK dimer was shown to be 1:1, suggesting that the allosteric site was symmetrically disposed to both subunits, and was perhaps located at the dimer interface.

The X-ray crystal structure of MK in the presence of mevalonate, AMPPNP and DPM showed that two molecules of DPM are bound, one to each subunit, in the mevalonate binding pocket (Fig. 3A) [12]. This result is perhaps not surprising, given that DPM is a partial bisubstrate analogue whose pyrophosphoryl moiety could take the place of the beta- and gamma-phosphates of ATP; however, the result stands at odds with the functional data described above and may represent a crystallographic artifact. The structure also revealed a pore at the dimer interface having excellent charge- and shape-complementarity to DPM; this pore could be the allosteric site. (Fig. 3B). We are currently pursuing the solution-phase structure of the fully-liganded MK (bound to mevalonate, AMPPNP and DPM) by NMR, to identify this site and study structural changes to the enzyme that occur upon allosteric binding.

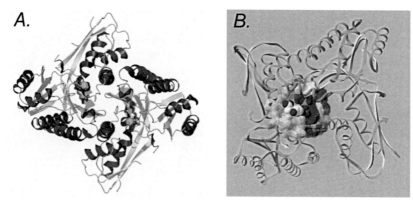

Figure 3. *Locating the DPM Allosteric Site.* (A) Crystal structure of *S. pneumoniae* MK with DPM (magenta sticks) and Mg^{2+} (green spheres) bound (PDB: 2O12). (B) MK model with DPM positioned at a pore in the subunit interface. Vacuum electrostatics near the pore are displayed as colored surfaces (blue, positive; red, negative).

PHOSPHOMEVALONATE KINASE

The X-ray structure of phosphomevalonate kinase presents a classic GHMP kinase scaffold. Comparison of the apo and ternary-complex forms of the enzyme (Fig. 4) reveals that four regions undergo significant conformational changes as a result of ligand binding [13, 14].

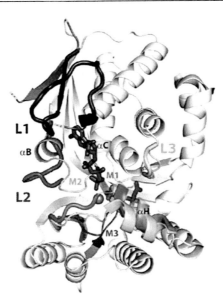

Figure 4. *Comparison of the Apo and Ternary-Complex Structure of PMK from S. pneumoniae.* Regions that do not change noticeably upon ligand binding are grey, the four responsive elements (L1, L2, L3 and αH) are colored in blue, red, cyan and green, respectively – the more intense colors are associated with the ternary complex.

These regions (L1, L2 L3 and αH) are color-coded, and the more intense colors are associated with the ternary complex. L1 (blue) is disordered in the absence of ligand, and reorganizes upon binding of nucleotide (AMPPNP, purple) with the result that residues that would otherwise obstruct binding are withdrawn from the binding pocket, and hydrogen bonds to the adenine ring are established. L2 (red) undergoes a considerable structural change in which a small helical element unravels to deliver the ammonium group of Lys101 into direct contact with the β,γ-bridging atom of AMPPNP (Fig. 5), where it will stabilize the negative charge expected to develop at this position during bond cleavage in a dissociative reaction. In the apo structure, Lys101 forms a salt bridge to the carboxylate of Glu98, and is positioned to obstruct the binding of nucleotide. Upon binding, the salt bridge is broken, the helix unwinds, "swinging" Lys101 past Lys100, which hydrogen bonds to the amide backbone of Lys208. The structural change resemble a "lysine switch" that when thrown, activates catalysis. It is interesting to note that while other GHMP kinases also feature a lysine switch, the catalytic lysine has migrated to a different position in the active site.

Lefurgy, S.T. and Leyh, S.

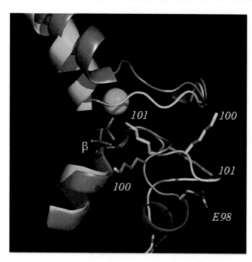

Figure 5. *A Di-lysine Catalytic Switch.* The L2 regions of the apo- (purple) and ternary-complex (cyan) forms of PMK are overlain. The β- and γ- phosphoryl groups of AMPPNP are shown.

The structure of the PMK ternary complex suggested that the substrates are positioned in a non-reactive orientation. The phosphoryl-group of phosphomevalonate is nearly orthogonal to what is expected for in-line nucleophillic attack at the γ-phosphate of ATP (Fig. 6A). Comparison of the PMK arrangement to that of six other GHMP-kinase ternary complexes revealed that of the seven total structures, three exhibited the non-reactive arrangement, three exhibited what appeared to be an excellent positioning for in-line attack (e. g., Fig. 5B), and one, erythritol kinase, exhibited both conformations. Clearly, these active sites can bind substrates in either conformation, suggesting that these forms might interconvert during the catalytic cycle. To assess the likelihood that the PMK active site might accommodate interconversion, the substrate configurations were aligned by superposing the C_α-traces of the ternary complex of PMK with that of Gal-NAc kinase – a good example of a reactive complex. The alignment suggested that the only large-scale movement was the migration of the divalent cation between "walls" of the active site, and that the active site was indeed capacious enough to accommodate such migration.

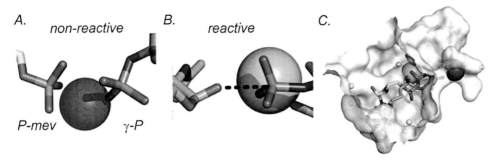

Figure 6. *Reactive and Non-Reactive Positioning of Substrates in the GHMP-Kinase Family.* (A) Positioning in the PMK ternary complex. (B) Positioning in the Gal-Nac kinase complex. (C) Superposition of the reactive and non reactive structures in the PMK active site. Color scheme: *reactive* position – grey nucleotide, green Mg^{+2}; *non-reactive* position – atom-colored nucleotide, red Mg^{2+}; aqua spheres – water.

The PMK active site harbors a water pentamer – five molecule of water arranged with an oxygen at each apex. Five of the ten pentamer-water hydrogens help form the ring (each shared by two oxygen atoms); the remaining five are distributed radially. The geometry of the pentamer allows the ring-hydrogen bonds to form with little if any strain [15], resulting in strong ring-hydrogen bonds and acidification of the radial protons. Four of the pentamer positions are occupied well in the apo-form of the enzyme, the fifth appears to be stabilized binding of ligands. The radial protons of the pentamer form hydrogen bonds to the negatively charged aspects of the substrates and appear well positioned to participate in chemistry (Fig. 7). Indeed, the reactive moieties of the substrates are interlaced in a large, complex water structure that remarkably manages to foster chemistry without short-circuiting it by hydrolyzing the γ-phosphate of ATP.

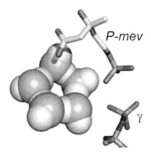

Figure 7. *Substrate Interacts with an Active-Site Water Pentamer.* The β- and γ-phosphates of AMPPNP are shown. Pentamer color scheme: orange – oxygen; white – protons.

Diphosphomevalonate Decarboxylase

The final step in the mevalonate pathway is carried out by diphosphomevalonate decarboxylase (DPM-DC), which phosphorylates the C_3-hydroxyl of DPM and subsequently eliminates phosphate and CO_2. This reaction is thought to proceed in a stepwise manner, with phosphate departing first, resulting in the formation of a carbocation. Abeles and coworkers provided evidence that a carbocation transiently forms by showing that substitution of the C_3-methyl group with a hydrogen or a fluoromethyl group causes the reaction to halt after phosphorylation, implying that the electron-donating methyl group is required to stabilize the carbocation inductively [16]. We sought to exploit carbocation formation to inactivate DPM-DC through use of DPM analogues in which the C_3-methyl group was replaced by substituents that were able, by resonance with the carbocation, to generate strongly electrophilic species that could become covalently bound to the enzyme (Fig. 8) [3]. One of these analogues contained a cyclopropyl substituent that was proposed to undergo ring opening upon ionization, forming a homoallyl cation that is stabilized by the loss of 27 kcal/mol of ring strain [17, 18]. If this rearrangement occurred, we anticipated that a nucleophile on the enzyme surface could attack the carbocationic intermediate, forming a covalent bond and inactivating the enzyme. However, inactivation did not occur over the course of >1000 turnovers, suggesting that no such adduct had formed. We therefore monitored the fate of the analogue during turnover using the unique NMR signal of the cyclopropyl protons. If the ring opened and formed a primary alcohol (by quenching with water), the proton chemical shifts associated with the ring would move significantly downfield [19]. Instead, we observed only a very small change (0.3 ppm) in the cyclopropyl proton chemical shifts that was consistent with an intact cyclopropyl group adjacent to a carbon-carbon double bond; this structure was subsequently confirmed by 2-D NMR. These data rule out a ring-opened product, and suggest that the analogue undergoes a decarboxylation that resembles that of the native substrate, DPM.

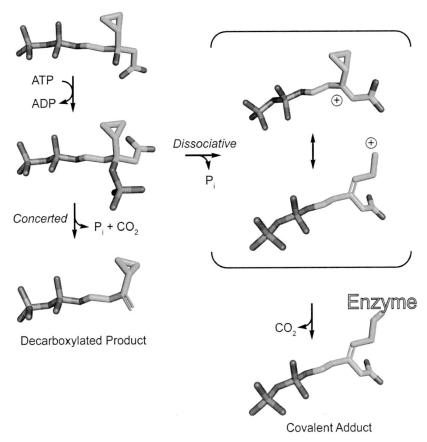

Figure 8. *Probing the Mechanism of DPM-DC with Cyclopropyl-DPM.* An analogue of DPM with a cyclopropyl substitution at C_3 is phosphorylated by the enzyme. If a carbocation forms during the reaction (*Dissociative* branch), it can be delocalized into the cyclopropyl ring, causing the ring to open, forming a strong electrophile. This ring-opened carbocation can be quenched by a nucleophile on the enzyme surface or water. Alternatively, a concerted mechanism produces only decarboxylated products.

These results call into question the extent of carbocation formation in the DPM-DC transition state. If a full carbocation formed, we expected to observe rearrangement of the cyclopropyl group. It is conceivable that a carbocation could form but not react, provided that no nucleophile is in close proximity to the carbocation (the protein surface or water); however, the presence of four crystallographic water molecules in the large pocket adjacent to C_3 and the proximity of the Asp276, Lys18, Ser185, and Met189 side chains and the Tyr19 carbonyl as potential nucleophiles argue against this possibility (Fig. 9). Alternatively, carbocation formation may be minimal or absent, in which case elimination of the carboxylate and the phosphate is concerted rather than dissociative. Abeles' work on the effects of altered electron induction at C_3 with the mammalian enzyme shows clearly that DPM-DC chemistry is sensitive to such changes and supports the development of a positive charge at

C_3 in the transition state [16]. Further, by replacing C_3 with a positively charged tertiary amine, he created an analogue that mimicked the structure and charge characteristics of a dissociative transition state. The affinity of this analogue (0.75 μM) was only 20-fold higher than that of the substrate [20], which, while supportive of a dissociative character in the transition state, is perhaps more consistent with development of partial rather than complete positive charge at C_3. While studies that correlate the extent of positive charge formation with degree of ring opening in cyclopropyl ring systems do not yet exist, our results, which demonstrate no detectible ring opening, are consistent with only slight positive charge formation in the transition state. Using kinetic isotope effects it may be possible to directly assess the extent of positive charge development on C_3 at the transition state.

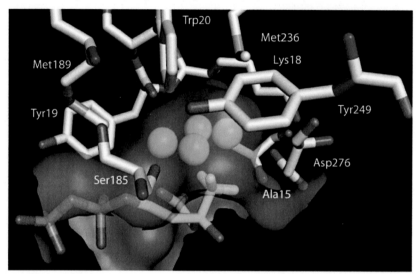

Figure 9. *DPM-DC Substrate Binding Pocket with Modeled DPM.* A homology model of the *S. pneumoniae* DPM-DC based on the 70% identical *Streptococcus pyogenes* DPM-DC (white, PDB: 2GS8) is shown with manually positioned DPM (cyan) (for full description, see [3]. The DPM C3-methyl points into a large water-filled cavity composed of nine conserved residues. The surface (light blue) represents the Van der Waals contact surface of the protein model.

OUTLOOK FOR MEVALONATE PATHWAY INHIBITION

Diphosphomevalonate interacts with each of enzymes in the *S. pneumoniae* mevalonate pathway – as an allosteric inhibitor of MK, a reaction product of PMK and a substrate for DPM-DC. Thus, non-reactive analogues of DPM have the potential to inhibit all three of the enzymes in this pathway. A challenge for moving DPM-based antibiotics into the clinic will be getting a highly charged (4-) molecule across the bacterial membrane. It should be possible to overcome this barrier using a prodrug strategy in which uncharged mevalono-lactone or mevalonate analogue esters are taken up by the cell, saponified and fed into the endogenous mevalonate pathway, which converts them to the active diphosphorylated species *in situ* [21, 22]. It must also be considered that, while DPM selectively inhibits pneumococcal MK, DPM analogues may cross-react with the human homologues of PMK and DPM-DC. However, inhibitors of the human mevalonate pathway, such as the widely-used cholesterol-lowering statins and anti-proliferative bisphosphonates, are generally very well tolerated and have shown the potential to have a clinical impact on asthma [23, 24], wound-healing [25], hepatitis [26, 27] and HIV [28], suggesting that even nonselective inhibitors have potential therapeutic benefits.

REFERENCES

[1]   Obaro, S., and Adegbola, R. (2002) The pneumococcus: carriage, disease and con-jugate vaccines. *J. Med. Microbiol.* **51**:98–104.

[2]   Schuchat, A., Robinson, K., Wenger, J.D., Harrison, L.H., Farley, M., Reingold, A.L., Lefkowitz, L., and Perkins, B.A. (1997) Bacterial meningitis in the United States in 1995. Active Surveillance Team. *N. Engl. J. Med.* **337**:970–976.

[3]   Lefurgy, S.T., Rodriguez, S.B., Park, C.S., Cahill, S., Silverman, R.B., and Leyh, T.S. Probing ligand-binding pockets of the mevalonate pathway enzymes from *Strepto-coccus pneumoniae. J. Biol. Chem. in press.*

[4]   Van Bambeke, F., Reinert, R.R., Appelbaum, P.C., Tulkens, P.M., and Peeter-mans, W.E. (2007) Multidrug-resistant *Streptococcus pneumoniae* infections: current and future therapeutic options. *Drugs* **67**:2355–2382.
        doi: http://dx.doi.org/10.2165/00003495-200767160-00005.

[5]   Kyaw, M.H., Lynfield, R., Schaffner, W., Craig, A.S., Hadler, J., Reingold, A., Thomas, A.R., Harrison, L.H., Bennett, N.M., Farley, M.M., Facklam, R.R., Jorgen-sen, J.H., Besser, J., Zell, E.R., Schuchat, A., and Whitney, C.G. (2006) Effect of introduction of the pneumococcal conjugate vaccine on drug-resistant *Streptococcus pneumoniae. N. Engl. J. Med.* **354**:1455–1463.
        doi: http://dx.doi.org/10.1056/NEJMoa051642.

[6]   Huang, S.S., Hinrichsen, V.L., Stevenson, A.E., Rifas-Shiman, S.L., Kleinman, K., Pelton, S.I., Lipsitch, M., Hanage, W.P., Lee, G.M., and Finkelstein, J.A. (2009) Continued impact of pneumococcal conjugate vaccine on carriage in young children. *Pediatrics* **124**:e1–11.
        doi: http://dx.doi.org/10.1542/peds.2008-3099

[7]   Lange, B.M., Rujan, T., Martin, W., and Croteau, R. (2000) Isoprenoid biosynthesis: The evolution of two ancient and distinct pathways across genome. *Proc. Natl. Acad. Sci. U.S.A.* **97**:13172–13177.
        doi: http://dx.doi.org/10.1073/pnas.240454797.

[8]   Wilding, E.I., Brown, J.R., Bryant, A.P., Chalker, A.F., Holmes, D.J., Ingraham, K.A., Iordanescu, S., So, C.Y., Rosenberg, M., and Gwynn, M.N. (2000) Identification, evolution, and essentiality of the mevalonate pathway for isopentenyl diphosphate biosynthesis in gram-positive cocci. *J. Bacteriol.* **182**:4319–4327. doi: http://dx.doi.org/10.1128/JB.182.15.4319-4327.2000.

[9]   Hinson, D.D., Chambliss, K.L., Toth, M.J., Tanaka, R.D., and Gibson, K.M. (1997) Post-translational regulation of mevalonate kinase by intermediates of the cholesterol and nonsterol isoprene biosynthetic pathways. *J. Lipid Res.* **38**:2216–2223.

[10]   Dorsey, J.K., and Porter, J.W. (1968) The inhibition of mevalonic kinase by geranyl and farnesyl pyrophosphates. *J. Biol. Chem.* **243**:4667–4670.

[11]   Andreassi, J.L., 2nd, Dabovic, K., and Leyh, T.S. (2004) *Streptococcus pneumoniae* isoprenoid biosynthesis is downregulated by diphosphomevalonate: an antimicrobial target. *Biochemistry* **43**:16461–16466. doi: http://dx.doi.org/10.1021/bi048075t.

[12]   Andreassi, J.L., 2nd, Bilder, P.W., Vetting, M.W., Roderick, S.L., and Leyh, T.S. (2007) Crystal structure of the *Streptococcus pneumoniae* mevalonate kinase in complex with diphosphomevalonate. *Protein Sci.* **16**:983–989. doi: http://dx.doi.org/10.1110/ps.072755707.

[13]   Andreassi, J.L., 2nd, Vetting, M.W., Bilder, P.W., Roderick, S.L., and Leyh, T.S. (2009) Structure of the ternary complex of phosphomevalonate kinase: the enzyme and its family. *Biochemistry* **48**:6461–6468. doi: http://dx.doi.org/10.1021/bi900537u.

[14]   Romanowski, M.J., Bonanno, J.B., and Burley, S.K. (2002) Crystal structure of the *Streptococcus pneumoniae* phosphomevalonate kinase, a member of the GHMP kinase superfamily. *Proteins* **47**:568–571. doi: http://dx.doi.org/10.1002/prot.10118.

[15]   Harker, H.A., Viant, M.R., Keutsch, F.N., Michael, E.A., McLaughlin, R.P., and Saykally, R.J. (2005) Water pentamer: characterization of the torsional-puckering manifold by terahertz VRT spectroscopy. *J. Phys. Chem. A* **109**:6483–6497. doi: http://dx.doi.org/10.1021/jp051504s.

[16]   Dhe-Paganon, S., Magrath, J., and Abeles, R.H. (1994) Mechanism of mevalonate pyrophosphate decarboxylase: evidence for a carbocationic transition state. *Biochemistry* **33**:13355–13362. doi: http://dx.doi.org/10.1021/bi00249a023.

[17]   Hart, H., and Sandri, J.M. (1959) The Solvolysis of *p*-Nitrobenzoates of Certain Cyclopropylcarbinols. *J. Am. Chem. Soc.* **81**:320–326. doi: http://dx.doi.org/10.1021/ja01511a016.

[18]   Liebman, J.F., and Greenberg, A. (1976) A Survey of Strained Organic Molecules. *Chem. Rev.* **76**:311–365.
doi: http://dx.doi.org/10.1021/cr60301a002.

[19]   Gunther, H. (1997) *NMR Spectroscopy: Basic Principles, Concepts, and Applications in Chemistry.* John Wiley Wolfenden, R. (1999) Conformational aspects of inhibitor design: enzyme-substrate interactions in the transition state. *Bioorg. Med. Chem.* 7:647–652.
doi: http://dx.doi.org/10.1016/S0968-0896(98)00247-8.

[21]   Cuthbert, J.A., and Lipsky, P.E. (1990) Inhibition by 6-fluoromevalonate demonstrates that mevalonate or one of the mevalonate phosphates is necessary for lymphocyte proliferation. *J. Biol. Chem.* **265**:18568–18575.

[22]   Cuthbert, J.A., and Lipsky, P.E. (1995) Suppression of the proliferation of Ras-transformed cells by fluoromevalonate, an inhibitor of mevalonate metabolism. *Cancer Res.* **55**:1732–1740.

[23]   Camoretti-Mercado, B. (2009) Targeting the airway smooth muscle for asthma treatment, *Transl. Res.* **154**:165–174.
doi: http://dx.doi.org/10.1016/j.trsl.2009.06.008.

[24]   Zeki, A.A., Franzi, L., Last, J., and Kenyon, N.J. (2009) Simvastatin inhibits airway hyperreactivity: implications for the mevalonate pathway and beyond. *Am. J. Respir. Crit. Care Med.* **180**:731–740.
doi: http://dx.doi.org/10.1164/rccm.200901-0018OC.

[25]   Vukelic, S., Stojadinovic, O., Pastar, I., Vouthounis, C., Krzyzanowska, A., Das, S., Samuels, H.H., and Tomic-Canic, M. (2009) Farnesyl pyrophosphate inhibits epithelialization and wound healing through the glucocorticoid receptor. *J. Biol. Chem.* **285**:1980–1988.
doi: http://dx.doi.org/10.1074/jbc.M109.016741.

[26]   Lyn, R.K., Kennedy, D.C., Sagan, S.M., Blais, D.R., Rouleau, Y., Pegoraro, A.F., Xie, X. S., Stolow, A., and Pezacki, J.P. (2009) Direct imaging of the disruption of hepatitis C virus replication complexes by inhibitors of lipid metabolism. *Virology* **394**:130–142.
doi: http://dx.doi.org/10.1016/j.virol.2009.08.022.

[27]   Ye, J., Wang, C., Sumpter, R., Jr., Brown, M.S., Goldstein, J.L., and Gale, M., Jr. (2003) Disruption of hepatitis C virus RNA replication through inhibition of host protein geranylgeranylation. *Proc. Natl. Acad. Sci. U.S.A.* **100**:15865–15870.
doi: http://dx.doi.org/10.1073/pnas.2237238100.

[28]   del Real, G., Jimenez-Baranda, S., Mira, E., Lacalle, R.A., Lucas, P., Gomez-Mouton, C., Alegret, M., Pena, J.M., Rodriguez-Zapata, M., Alvarez-Mon, M., Martinez, A.C., and Manes, S. (2004) Statins inhibit HIV-1 infection by down-regulating Rho activity. *J. Exp. Med.* **200**:541–547.
doi: http://dx.doi.org/10.1084/jem.20040061

Estrogen Sulfotransferase – a Half-Site Reactive Enzyme

Thomas S. Leyh

The Department of Microbiology and Immunology,
The Albert Einstein College of Medicine,
1300 Morris Park Ave., Bronx, NY 10461, U.S.A.

E-Mail: thomas.leyh@einstein.yu.edu

Received: 17th June 2010 / Published: 14th September 2010

Abstract

Estrogen sulfotransferase (EST) regulates the biological activity of
estradiol – sulfation of this important signalling molecule inactivates
its estrogen-receptor-binding activity. EST is a cytosolic sulfotransferase – one of family of structurally well conserved enzymes that exhibit
broad, overlapping substrate specificities and that together comprise a
robust catalytic network designed to handle the many substrates present in the cell. We have discovered recently that EST is a half-site
enzyme – only one of the subunits of the dimer is capable of forming
product during turnover. Thus, during the catalytic cycle a molecular
"decision" is made that couples the silencing one of the subunits to the
activity of the other. The discovery of this remarkable behaviour and
the implications of the EST mechanism for the sulfotransferase field
are the subject of this article.

INTRODUCTION

Estrogen sulfotransferase (EC 2.8.2.4), a dimer of identical 35 kDa protomers, catalyzes the transfer of the sulfuryl moiety ($-SO_3$) of PAPS (3'-phosphoadenosine 5'-phosphosulfate) to the 3'-hydroxyl of 17β-estradiol (E_2), to form E_2S (mechanism I).

$$E_2 + PAPS^{4-} \rightleftharpoons E_2S^- + PAP^{4-} + H^+ \tag{I}$$

The presence of the sulfuryl-group prevents estradiol from binding the estrogen receptor at physiologically relevant concentrations [1]. The group is removed hydrolytically by sulfatases, which regenerate E_2, and the coordinated actions of the sulfotransferase and sulfatase balance the sulfated and non-sulfated forms of E_2 to meet the metabolic requirements of the cell.

It is generally agreed that sulfation/desulfation cycles are responsible for regulating a large, diverse set of metabolites and cellular processes including steroid hormones [2, 3], signalling peptides [4 – 7], heparin [8, 9], haemostasis [10 – 12], lymph circulation [13] and numerous drugs and xenobiotics [14, 15].

Approximately ten distinct sulfotransferases have been identified in the cytosols of mammalian cells [16, 17]. These enzymes typically exhibit a broad substrate specificity that is centred on particular molecular traits of non-nucleotidyl substrate (or acceptor), and the enzyme common names reference the metabolite for which the enzyme is believed to exhibit its highest affinity. The substrates specificities of these enzymes often overlap, producing a redundant and robust metabolic network designed to manage the sulfation requirements of the organism.

A SUGGESTION OF HALF-SITE REACTIVITY

In characterizing the rates and equilibria of reactions occurring on the surface of the recombinant, human estrogen sulfotransferase (EST) during its catalytic cycle, we discovered that the enzyme produces a presteady-state burst of the product [18]. Such behaviour reveals that a rate-determining step(s) occurs after the product has formed on the enzyme; however, it gives no indication as to where in the product release branch of the catalytic cycle that step(s) occurs. The amplitude of the burst corresponds to precisely one equivalent of product formed per dimer, suggesting for the first time that the enzyme might be half-site reactive (Fig. 1).

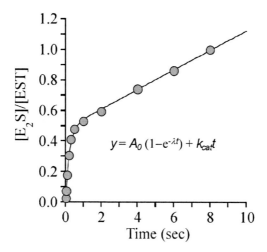

Figure 1. A burst of E_2S formation. E_2S synthesis was initiated by rapidly mixing a solution containing EST (1.0 µM, dimer), [^{3}H]E_2 (3.6 µM, 720 x K_m, SA = 90 Ci/mmol), glycerol (10% v/v), $MgCl_2$ (7.0 mM), DTT (1.0 mM), and 50 mM KPO_4 (pH 6.3) with a solution of equal volume that was identical except that it lacked EST and E_2, and contained PAPS (18.0 µM, 305 x K_m). Reactions were quenched by rapidly mixing the reacting solutions with an equal volume of HCl (0.66 M). [^{3}H]E_2 was extracted from the quenched mixture using CCl_4 and [^{3}H]E_2S, which remained in the aqueous phase, was counted. All solutions were equilibrated at 25 (±2) °C prior to mixing. Reactions were performed in triplicate and averaged. The smooth curve represents the best-fit of the averaged data to the equation: $A_0 (1-e^{-\lambda t}) + k_{cat}t$.

Half-site reactivity is a curious phenomenon in which seeming identical subunits are locked into a relationship in which only one out of every two subunits of a multi-subunit enzyme is capable of forming product. Thus, at some point in the catalytic cycle, a "molecular decision" is made to assign reactivity to a specific subunit – a decision almost certainly rooted in a structural asymmetry that allosterically links the behaviours of the two active sites to one another.

CONFIRMATION OF HALF-SITE REACTIVITY

While the observed burst amplitude is suggestive of half-site behaviour, it is equally consistent with a mechanism in which, for example, the equilibrium constant governing interconversion of the central complexes of the enzyme (the *internal equilibrium constant*, or $K_{eq\ int}$) is equal to 1, and the release of product is sufficiently slow to allow the interconversion to approach equilibrium.

A burst amplitude provides a stoichiometry, i.e. [product formed]/[active sites occupied]. Assuming a half-site mechanism, the value of 0.52/dimer requires that $K_{eq\ int}$ be considerably larger than one (≥ ~10) on the subunit that turns over – where this is not the case, the amplitude would be less. The value of 0.52 was determined under conditions where both active sites of the dimer are saturated. Under conditions in which only one of the subunits of the dimer has substrate bound, virtually all of the bound substrate is expected to be converted to the product due to the favourable internal energetics of the active subunit, and the stoichiometry (normalized to bound active sites) will approach ~1.0. On the contrary, the full-site model predicts a stoichiometry of 0.5 regardless of the occupancy of dimer active sites (due to the restriction that $K_{eq\ int} = 1$). Thus, the mechanisms can be distinguished on this basis.

Capitalizing on these differences to ascertain the operative mechanism, the stoichiometry was determined as a function of occupancy of the EST dimer. Occupancy was controlled by varying the $[EST]_{active\ site}/[E_2]$ ratio at a fixed concentration of E_2 (1.0 µM) – PAPS is fixed and saturating in all cases. At the lowest [EST] (1.0 µM), the majority of dimers have both active sites occupied, and either mechanism predicts an amplitude of 0.5 (which is what is observed, Figure 2). At the highest [EST] (12 µM), simple statistics (i.e., independent subunits) predict that 17% of the dimers will have both of their active sites occupied; 83% will have only one. At this condition, the half-site mechanism yields a maximum amplitude of ~91%; whereas, the full-site model yields 5%. Figure 2 reveals how the amplitude varies as a function of [EST], and compares the experimental data (▲) to the result predicted for the half-site (●) and full-site (◉) mechanisms. The experimental data follow the predictions of the half-site model well, and differ sharply from the expectations of the full-site mechanism. Thus it appears that EST is a half-site enzyme.

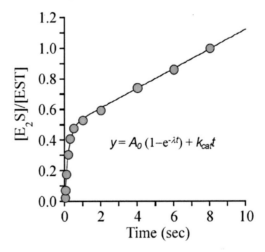

Figure 2. Confirmation of half-sites reactivity. Reactions were initiated by mixing a solution containing PAPS (16 µM), glycerol (10% v/v), $MgCl_2$ (7.0 mM), DTT (1.0 mM), and 50 mM KPO_4 (pH 6.3) with an equal volume of an identical solution that did not contain PAPS, but did contain $[^3H]E_2$ (2.0 µM) and various concentrations of EST (2.0, 3.0, 4.0, 6.0, 8.0, 16.0, 24.0 µM – monomer concentrations). The plotted E_2S concentrations represent the E_2S formed in reactions quenched at 0.30 s – at this

time-point, the burst is essentially complete and steady-state turnover contributes only slightly to overall product formation (see Figure 1). Triangles ($\blacktriangle$) represent experimentally determined [E$_2$S]; red circles ($\bullet$) represent [E$_2$S] predicted using a half-site reactivity model in which $K_{eq\ int} \gg 1.0$; tan circles ($\circ$) represent [E$_2$S] predicted using a full-site reactivity model with $K_{eq\ int} \sim 1.0$. Experimental values were determined in triplicate and averaged. All solutions were equilibrated at 25 (± 2) °C prior to mixing.

An important question in the sulfotransferase field is how, confronted with the considerable array of substrate options provided by the cell, a given sulfotransferase selects one substrate over another. Given a simple mechanism in which substrate (S) binds to enzyme (E) and is converted to a product that is released slowly enough to allow both binding and the ES – EP interconversion to approach equilibrium, K_m is approximated by the product of the equilibria for these two steps. Such mechanisms provide a flexible, two-stage selection process that offers certain benefits to the cell. For example, it offers the opportunity to provide tight overall binding while maintaining binding at the first step that is sufficiently weak to allow the active site to sample numerous substrates during selection.

SELECTIVITY AND THE EST MECHANISM

Notwithstanding electrostatic guidance [19], the maximum microscopic rate constant for the binding of ligand to an enzyme active site is given by the diffusion on-rate constant, $\sim 8 \times 10^8$ M^{-1}s^{-1} [20]. However, because binding is a complex process involving multiple, sequential events to achieve docking of ligand, the docking on-rate constant is often one-to-two orders of magnitude less that the diffusion rate constant. K_m for estradiol is 5 nM. If the affinity of EST for E$_2$ were determined solely at the binding step, the maximum k_{off} (assuming diffusion-limited k_{on}) can be estimated at ~ 4 s^{-1}, and the half-life of E$_2$ on the surface of EST is ~ 0.17 s^{-1}. If k_{on} were one-to-two orders of magnitude less than diffusion, the half-life increases to 17 s. On the other hand, if the affinity is determined in two steps, where the internal equilibrium constant is, say, 50 in favour of product, then the half-lives will range, instead, from $0.0035 - 0.35$ s^{-1}. Hence, addition of the second step allows the system to maintain the overall affinity for substrate while allowing to enzyme to sample far greater number of substrates in a given period of time.

The active-site surface of an enzyme restructures continuously as the system moves through its catalytic cycle – R-groups are recruited and dispatched to perform reaction-stage specific functions. Whether a particular enzyme form participates in substrate selection is determined by whether it occurs before or after the system has committed to the forward reaction. For example, in the case where a single step is rate-determining, all forms that appear subsequent to the slow step will not contribute to selection, since the system cannot return from that region of the reaction coordinate – it is committed to the forward path. Thus, as product release becomes slow enough to allow both binding and surface chemistry to approach equilibrium, the product-bound forms of the enzyme become active in the selection of substrate, and $K_{eq\ int}$ can become a significant selection determinant.

While product release from EST is not slow enough to allow both binding and chemistry to reach near-equilibrium, it *is* slow enough to approach this condition and for the effects of the internal equilibrium constant to play significantly into the selection process. To appreciate how the EST mechanism can produce such effects, consider Mechanism II, which represents the binding PAPS to $E \cdot E_2$, followed by formation and release of products (PAPS = S, $E \cdot E_2 = E$).

$$E \underset{k_{off}}{\overset{[S]k_{on}}{\rightleftharpoons}} E \cdot S \underset{k_2}{\overset{k_1}{\rightleftharpoons}} E \cdot P \overset{k_3}{\rightarrow} E + P \tag{II}$$

$K_{m\ S}$ for this mechanism is given by Equation 1.

$$K_m = \frac{k_{off} k_2 + k_{off} k_2 + k_1 k_3}{k_{on}(k_1 + k_2 + k_3)} \tag{1}$$

For the EST mechanism, it has been demonstrated that $k_1 \gg k_3$, $k_1 \gg k_2$, and $k_{off} \gg k_1$. Given these constraints, Equation 1 reduces to 2.

$$K_m = \frac{k_{off}}{k_{on}} \cdot \frac{k_2 + k_3}{k_1} \tag{2}$$

It is important to realize that $(k_2 + k_3)/k_1$ represents the ratio of [E·S] to [E·P] in the steady state, $([E \cdot S]/[E \cdot P])_{ss}$. That this is so can be seen by applying the steady-state assumption to the rate of formation and disappearance of EP:

$$\frac{dEP}{dt} = [ES]k_1 = [EP](k_2 + k_3) = -\frac{dEP}{dt} \tag{3}$$

Rearranging,

$$\left(\frac{[ES]}{[EP]}\right)_{SS} = \frac{k_2 + k_3}{k_1} \tag{4}$$

Thus, for the EST mechanism, K_m is a linear function of the steady-state, mass-ratio of the central complexes (Equation 2). Given experimental values for k_1, k_2 and k_3 (4.4 s^{-1}, 0.070 s^{-1} and 0.072 s^{-1} [18]), $([E \cdot S]/[E \cdot P])_{ss}$ is calculated at 0.031, which, when multiplied by K_d for PAPS binding to $E \cdot E_2$, yields a K_m of 130 nM – a value that is in reasonable agreement with the published value of 59 nM [21].

The EST mechanism has evolved to a point where its rate constants are balanced to control the steady-state central complex mass ratio, with the result that the affinity of both E_2 and PAPS are enhanced ~30-fold. It is interesting to consider that changes in this balance across a series of substrates will bias substrate selection, and further, that both the substrate and product forms of the acceptor are "scrutinized" by the enzyme during the selection phase of the catalytic cycle. One is left to wonder to what extent similar mechanisms play important roles in the selection of substrate within the highly conserved sulfotransferase family.

REFERENCES

[1]   Hahnel, R., Twaddle, E., and Ratajczak, T. (1973) The specificity of the estrogen receptor of human uterus. *J. Steroid Biochem.* **4**:21 – 31.
doi: http://dx.doi.org/10.1016/0022-4731(73)90076-9.

[2]   Falany, C.N., Wheeler, J., Oh, T.S., and Falany, J.L. (1994) Steroid sulfation by expressed human cytosolic sulfotransferases. *J. Steroid Biochem. Mol. Biol.* **48**: 369 – 375.
doi: http://dx.doi.org/10.1016/0960-0760(94)90077-9.

[3]   Pasqualini, J.R., Schatz, B., Varin, C., and Nguyen, B.L. (1992) Recent data on estrogen sulfatases and sulfotransferases activities in human breast cancer. *J. Steroid Biochem. Mol. Biol.* **41**:323 – 329.
doi: http://dx.doi.org/10.1016/0960-0760(92)90358-P.

[4]   Brand, S.J., Andersen, B.N., and Rehfeld, J.F. (1984) Complete tyrosine-O-sulphation of gastrin in neonatal rat pancreas. *Nature* **309**:456 – 458.
doi: http://dx.doi.org/10.1038/309456a0.

[5]   Jen, C.H., Moore, K.L., and Leary, J.A. (2009) Pattern and temporal sequence of sulfation of CCR5 N-terminal peptides by tyrosylprotein sulfotransferase-2: an assessment of the effects of N-terminal residues. *Biochemistry* **48**:5332 – 5338.
doi: http://dx.doi.org/10.1021/bi900285c.

[6]   Lee, S.W., Han, S.W., Sririyanum, M., Park, C.J., Seo, Y.S., and Ronald, P.C. (2009) A type I-secreted, sulfated peptide triggers XA21-mediated innate immunity. *Science* **326**:850 – 853.
doi: http://dx.doi.org/10.1126/science.1173438.

[7]   Roth, J.A., and Rivett, A.J. (1982) Does sulfate conjugation contribute to the metabolic inactivation of catecholamines in humans? *Biochem. Pharmacol.* **31**:3017 – 3021.
doi: http://dx.doi.org/10.1016/0006-2952(82)90073-9.

[8]   Hassan, H.H. (2007) Chemistry and biology of heparin mimetics that bind to fibroblast growth factors. *Mini Rev. Med. Chem.* **7**:1206 – 1235.
doi: http://dx.doi.org/10.2174/138955707782795665.

[9]   Lindahl, U. (2007) Heparan sulfate-protein interactions – a concept for drug design? *Thromb. Haemost.* **98**:109 – 115.

[10]   Atha, D.H., Lormeau, J.C., Petitou, M., Rosenberg, R.D., and Choay, J. (1985) Contribution of monosaccharide residues in heparin binding to antithrombin III. *Biochemistry* **24**:6723 – 6729.
doi: http://dx.doi.org/10.1021/bi00344a063.

[11]   Leyte, A., van Schijndel, H.B., Niehrs, C., Huttner, W.B., Verbeet, M.P., Mertens, K., and van Mourik, J.A. (1991) Sulfation of Tyr1680 of human blood coagulation factor VIII is essential for the interaction of factor VIII with von Willebrand factor. *J. Biol. Chem.* **266**:740 – 746.

[12]   Stone, S.R., and Hofsteenge, J. (1986) Kinetics of the inhibition of thrombin by hirudin. *Biochemistry* **25**:4622 – 4628.
doi: http://dx.doi.org/10.1021/bi00364a025.

[13]   Hortin, G.L., Farries, T.C., Graham, J.P., and Atkinson, J.P. (1989) Sulfation of tyrosine residues increases activity of the fourth component of complement. *Proc. Natl. Acad. Sci. U.S.A.* **86**:1338 – 1342.
doi: http://dx.doi.org/10.1073/pnas.86.4.1338.

[14]   Glatt, H. (2000) Sulfotransferases in the bioactivation of xenobiotics. *Chem. Biol. Interact.* **129**:141 – 170.
doi: http://dx.doi.org/10.1016/S0009-2797(00)00202-7.

[15]   Kauffman, F.C. (2004) Sulfonation in pharmacology and toxicology. *Drug Metab. Rev.* **36**:823 – 843.
doi: http://dx.doi.org/10.1081/DMR-200033496.

[16]   Coughtrie, M.W. (2002) Sulfation through the looking glass – recent advances in sulfotransferase research for the curious. *Pharmacogenomics J.* **2**:297 – 308.
doi: http://dx.doi.org/10.1038/sj.tpj.6500117.

[17]   Weinshilboum, R.M., Otterness, D.M., Aksoy, I.A., Wood, T.C., Her, C., and Raftogianis, R.B. (1997) Sulfation and sulfotransferases 1: Sulfotransferase molecular biology: cDNAs and genes. *FASEB J.* **11**:3 – 14.

[18]   Sun, M., and Leyh, T.S. (2010) The human estrogen sulfotransferase: a half-site reactive enzyme. *Biochemistry* **49**:4779 – 4785.
doi: http://dx.doi.org/10.1021/bi902190r.

[19]   Lu, B., and McCammon, J.A. (2010) Kinetics of diffusion-controlled enzymatic reactions with charged substrates. *PMC Biophys.* **3**:1.
doi: http://dx.doi.org/10.1186/1757-5036-3-1.

[20]   Albery, W.J., and Knowles, J.R. (1976) Evolution of enzyme function and the development of catalytic efficiency. *Biochemistry* **15**:5631 – 5640.
doi: http://dx.doi.org/10.1021/bi00670a032.

[21]   Zhang, H., Varlamova, O., Vargas, F.M., Falany, C.N., and Leyh, T.S. (1998) Sulfuryl transfer: the catalytic mechanism of human estrogen sulfotransferase. *J. Biol. Chem.* **273**:10888 – 10892.
doi: http://dx.doi.org/10.1074/jbc.273.18.10888.

Evolution of New Specificities in a Superfamily of Phosphatases

Karen N. Allen[1,*] and Debra Dunaway-Mariano[2,#]

[1]Department of Chemistry, Boston University, 590 Commonwealth Avenue,
Boston, MA 02215 – 2521, U.S.A.

[2]Department of Chemistry and Chemical Biology, University of New Mexico,
Albuquerque, NM, 87131, U.S.A.

E-Mail: *drkallen@bu.edu and #dd39@unm.edu

Received: 13th July 2010 / Published: 14th September 2010

Abstract

The evolution of new catalytic activities and specificities within an
enzyme superfamily requires the exploration of sequence space for
adaptation to a new substrate with retention of those elements required
to stabilize key intermediates/transition states as well as the core
enzyme fold. Phylogenetic analysis, mechanistic information, and
structure determination are used to reveal novel ways in which the
catalytic scaffold of a mechanistically diverse superfamily, the haloalkanoic acid dehalogenase enzyme superfamily, is tailored to new
biochemical functions. Newly uncovered substrate specificities and
activities in members of the superfamily are highlighted to explore
the interplay of function and form. We provide evidence that core
residues in this large enzyme family, form a "mold" in which the
trigonal bipyramidal transition-states formed during phosphoryl transfer are stabilized by electrostatic forces.

Introduction

The Haloalkanoic Acid Dehalogenase Superfamily (HADSF) is a large superfamily of
enzymes [1] with over 30,000 members to date [2]. Members are found in all three king-
doms of life, and are often observed several times within a given organism's genome (*e.g.* 29
in *E. coli* and 158 in humans) [3]. Although named for the first enzyme in the family to be

well characterized, the HADSF is primarily comprised of phosphohydrolases, with < 1% of members catalyzing dehalogenation reactions [4]. These minority reactions at carbon centers include 2-L-haloalkanoic acid dehalogenase (C-Cl cleavage) [5] and azetidine hydrolase (C-N cleavage) [6] and the more prevalent phosphoryl transfer reactions include the phosphonoacetaldehyde hydrolase (C-P cleavage) [7–8], Ca^{2+}-ATPase (PO-P cleavage) [1], phosphoserine phosphatase and phosphomannomutase (CO-P) [9–10] (Figure 1). Within this panoply of reactions, HADSF members act on a wide variety of substrates varying in steric bulk, electrostatics, and polarity. Here, we address how the phosphoryl hydrolases in this enzyme family have adapted to work on varying substrates while retaining high catalytic efficiency.

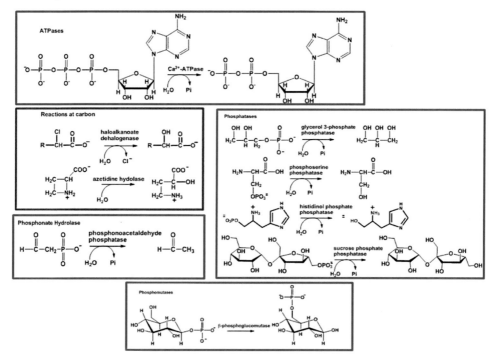

Figure 1. Representative reactions catalyzed by the HADSF

CONSERVED CHEMICAL AND CATALYTIC MECHANISM

Examination of the sequence similarity in the superfamily uncovers high diversity in primary structure, with only 10–15% sequence identity between homologues and closer to 30–40% identity within the strictly conserved catalytic motifs used to identify HADSF members [11]. These four motifs are positioned on loops at the C-terminal end of the parallel beta strands of the core Rossmann fold common to all superfamily members (Figure 2). Within motif 1 resides both the catalytic Asp nucleophile [9] and the general acid/base Asp catalyst which are integral to the mechanism of phosphoryl transfer [12 – 13]. The chemical mechanism of phosphoryl transfer itself is retained throughout the superfamily, and there is evidence that the catalytic mechanism is also retained.

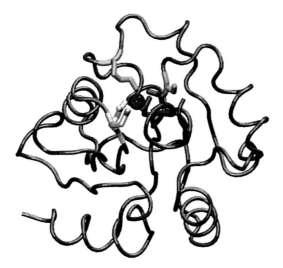

Figure 2. The conserved catalytic motifs of the HADSF phosphotransferases mapped onto the Rossmann fold catalytic scaffold (PDB 1LVH with cap domain removed for view into active site). Motif 1 (DXD) is depicted in red, motif 2 (S/T) in green, motif 3 (K/R) in cyan and motif 4 (DD, GDxxxD, or GDxxxxD) in yellow. The Mg^{2+} cofactor is depicted as a magenta sphere.

The commonality of the catalytic mechanism is supported by ultra-high resolution X-ray structures of transition-state analogues liganded to the enzyme hexose phosphate phosphatase (HPP) BT4131 from *Bacteroides thetaiotaomicron* VPI-5482 [13]. The complex of vanadate with HPP determined at 1.00 Å resolution assumes a trigonal bipyramidal coordination geometry with the nucleophilic Asp8 and one oxygen ligand at the apical position (Figure 3). Notably, the tungstate complex (1.03 Å resolution) assumes the same coordination geometry. The general acid/base residue Asp10 is critical to the stabilization of the trigonal bipyramidal species as evidenced by the collapse of the trigonal bipyramidal geometry in complexes of the Asp10Ala mutant complexed with vanadate (1.52 Å resolution) as well as tungstate (1.07 Å resolution). The ease of attaining the trigonal bipyramidal geometry for a given complex parallels the inhibition constants with K_i values for vanadate, tungstate and phosphate equal to 510 nM, 65 μM, and 5.2 mM, respectively. The core Rossmann fold stabilizes this trigonal bipyramidal transition state by engaging in favorable electrostatic interactions with the axial and equatorial atoms of the transferring phosphoryl group. Conserved backbone and side-chain interactions contributed by this scaffold were uncovered by a structural analysis of twelve liganded HADSF structures deposited in the protein data bank. Overall, these findings support the model that core domain residues in the

HADSF form a "mold" in which the trigonal bipyramidal transition states formed during phosphoryl transfer are stabilized by electrostatic forces, retaining a single catalytic mechanism for phosphoryl transfer.

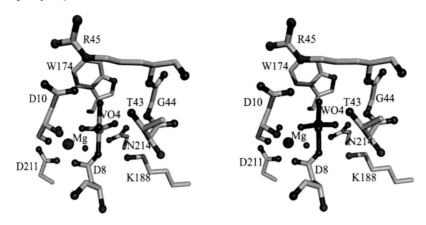

Figure 3. The trigonal bipyramidal complexes of wild type hexose phosphate phosphatase BT4131 from *Bacteroides thetaiotaomicron* VPI-5482 with vanadate (left) and tungstate (right) depicted as ball and stick and Mg^{2+} cofactor (magenta sphere).

THE CAP CONFERS SUBSTRATE SPECIFICITY

If the phosphoryl transfer mechanism is common among family members, how then is specificity conferred? The positioning of the four catalytic motifs provides an open active site that can be accessed by macromolecules (in enzymes such as magnesium dependent phosphatase 1 [14], T4 polynucleotide kinase/phosphatase [15], and RNA polymerase II C-terminal domain phosphatases [16], defined as C0 members) or adapted to small molecules using a cap domain. Thus the HADSF has a "modular" design, with the phosphoryl binding site conferred by the core domain being spatially distinct from the leaving-group binding site conferred by the cap domain. By far the most common cap assemblage (> 65%, Chetanya Pandya, unpublished results) in the HADSF is the C1 type cap wherein an alpha helical domain, which can vary in size from two helices to more than eight helices, is inserted between motifs 1 and 2. Analysis of these C1 members shows that the residues interacting with the leaving group originate from a single substrate specificity loop (Figure 4) [17]. The C2 cap is the second most populated type of cap assemblage, which is further divided into C2a and C2b subfamilies as defined by the topology of the α/β fold prototypical of these members. Depending on the size of the cap domain, there are up to two substrate specificity loops in C2 HADSF members [18 – 19].

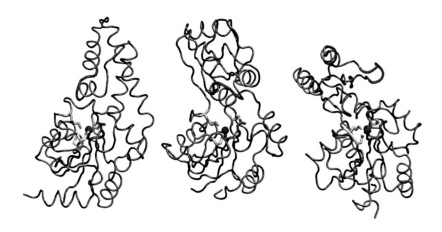

Figure 4. The substrate specificity loops (magenta) of type C1 (PDB 1LVH, left), C2a (PDB 2C4N, center) and C2b (PDB 1TJ5, right) HADSF cap assemblages. The catalytic motifs are colored as in Figure 2.

In addition to affording specificity residues in a single module, the dynamic movement of the cap domain relative to the core domain allows for ligand-induced cap closure [20] and, while giving access to the active site in the cap-open form, provides a solvent exclusive environment in the cap-closed form. In the cap closed form, the enzyme encapsulates the substrate such that the cavity remaining approximates the size and shape of the substrate. This feature has in fact been utilized to predict substrates in HADSF enzymes of unknown function [18]. The nature of cap closure differs between C1 and C2 caps, though in both cases the cap and core move as rigid bodies on two flexible linkers. The C1 caps move on a mechanical hinge (similar to the opening and closing of a clam) while the C2 caps close with a screw-like motion of the cap over the core. Thus, the cap domain is seemingly integral to specificity and chemical environment of the phosphoryl transfer reactions of HADSF members acting on small substrates.

PREDICTING SPECIFICITY FROM SEQUENCE

If this is the case, one might envision that examination of sequence alone might be used to predict if a cap exists and thus whether a given member acts on a small molecule substrate (C1 or C2 member) or on a phosphorylated protein (a capless, C0 member). Unfortunately, such a model is an oversimplification. The lack of a cap domain does not predict an open active site because oligomerization of the core domain can provide both specificity residues from an adjacent core domain as well as encapsulation of the active site environment (Figure 5). A prototypical example of such a C0 cap member is 2-keto-3-deoxy-D-gly-cero-D-galacto-9-phosphononic acid phosphate (KDN9P) phosphatase from *Bacteriodes thetaiotaomicron*, which acts in the biosynthetic pathway of the 9-carbon alpha-keto acid 2-keto-3-deoxy-D-glycero-D-galactononic acid [21]. Such polyhydroxylated α-keto acids are incorporated into cell-surface glycoproteins and glycolipids in both prokaryotic and eukaryotic organisms. The structure of KDN9P phosphatase complexed with the sialic acid N-acetyl neuraminate and vanadate to 1.63 Å resolution reveals a structure in which a small insert in the same position as the C1 cap domain allows tetramerization with active sites positioned at the subunit-subunit interface [21]. Indeed, seven out of the ten interactions made directly or through water between enzyme and the N-acetyl-neuraminic acid leaving group are made by the adjacent protomer core acting as a cap. The efficiency of this type of active site construction is demonstrated by the activity against the physiological substrate KDN9P ($K_m = 0.10$ mM, $k_{cat} = 1.2$ s^{-1}, $k_{cat}/K_m = 1 \times 10^4$ M^{-1}s^{-1}). Moreover, the substantively lower second-order rate constants against other related sugars such as the 8-carbon acid, 2-keto-3-deoxy-D-manno-8-phospho-octulosonic acid ($k_{cat}/K_m = 2 \times 10^2$ M^{-1}s^{-1}) shows the ability of the adjacent core residues to confer substrate discrimination. Notably, the tetrameric assembly adopted by KDN9P phosphatase does not appear to undergo significant substrate-induced rearrangement, unlike the typical cap-core movement observed in HADSF C1 and C2 phosphatases. Such motion is therefore not a requirement of substrate specificity or catalytic efficiency in the HADSF.

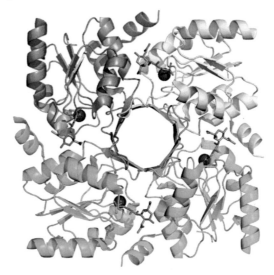

Figure 5. The structure of KDN9P phosphatase depicted as ribbons (colored by subunit) complexed with Mg^{2+} (magenta sphere), N-acetyl neuraminate and vanadate (grey sticks).

The adoption of a large cap domain to acquire substrate specificity elements versus a small insert to allow oligomerization (and use of the core for this same purpose) are not the sole mechanisms used by the HADSF to bind phosphate monoesters. The structure of D-*Glycero-*D-*manno*-heptose-1,7-(bis)phosphate phosphatase (GmhB) from *E. coli* bound to the substrate D-*Glycero*-D-*manno*-heptose-1,7-(bis)phosphate to 2.2 Å resolution reveals that, in place of a cap domain, the GmhB catalytic site is elaborated by three peptide inserts or loops in the core domain of each GmhB monomer that pack to form a continuous binding surface around the substrate leaving group [22] (Figure 6). Examination of the solvent-accessible surface of the substrate-liganded structure shows that the substrate leaving group forms a "plug" that occludes the passage between bulk solvent and the phosphoryl-transfer site. Together, the three inserts and bulky substrate leaving group shield the catalytic site from solvent. The question then arises, are the inserted segments sufficient to afford substrate specificity and catalytic efficiency?

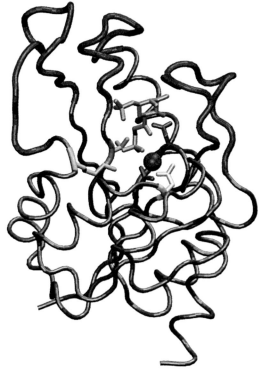

Figure 6. The structure of GmhB depicted as ribbons complexed with the substrate β-D-glycero-D-manno-heptose-1,7-(bis)phosphate (gold sticks). The substrate specificity loops inserted into the Rossmann core domain are shown in magenta. Motifs colored as in Figure 2.

This query is best approached by inspecting the steady-state kinetic properties of the enzyme. First, *E. coli* GmhB is highly efficient against its natural substrate β-D-glycero-D-manno-heptose 1,7-bisphosphate with $k_{cat}/K_m = 7 \times 10^6$ M^{-1}s^{-1}. Notably, orthologues of GmhB are utilized in the pathways for production of D-glycero-D-mannoheptose 1α-GDP [23] and L-glycero-D-manno-heptose 1β-ADP [24]. The pathways provide activated glycero-manno-heptose units for incorporation into membrane surface glycoproteins and glycolipids [25–29]. GmhB orthologues have been honed for hydrolysis of the anomer corresponding to that of the respective pathway kinase that generates the bis-phosphorylated substrate.

Consequently, *E. coli* GmhB shows a 100:1 preference for the β-anomer over the α-anomer while the *Bacteriodes thetaiotaomicron* GmhB has a 5-fold preference for the α-anomer over the β-anomer [30]. Site-directed mutagenesis demonstrates that conserved residues in the three inserted segments are integral to substrate binding and that the contribution to binding and activity differs between anomers [22]. Overall, the inserted segments are necessary and sufficient to confer anomeric selectivity, substrate specificity, and catalytic efficiency.

CONCLUSION

The HADSF is distinctive among enzyme superfamilies in that the elements responsible for the conserved chemistry are structurally and spatially distinct from those responsible for substrate specificity. Interactions between the core domain amino-acid side chains, main chain, and the substrate phosphoryl group make a "mold" for the trigonal bipyramidal transition state common to the phosphoryl transferases in the superfamily. Despite this conserved catalytic mechanism, a wide variety of substrates is allowed by utilizing determinants to bind the substrate leaving group, provided by inserted domains, loop extensions or segments that allow oligomerization and utilization of adjacent protomers. This modular approach to binding of new substrates is proposed to underlie successful evolution in the HADSF.

ACKNOWLEDGEMENT

This work was supported by NIH GM061099 (to DD-M and KNA).

REFERENCES

[1]   Collet, J.F., van Schaftingen, E., and Stroobant, V. (1998) A new family of phospho-transferases related to P-type ATPases. *Trends Biochem. Sci.* **23**(8):284.
       doi: http://dx.doi.org/10.1016/S0968-0004(98)01252-3.

[2]   Bateman, A. *et al.* (2004) The Pfam protein families database. *Nucleic Acids Res.* **32**(Database issue):D 138 – 41.
       doi: http://dx.doi.org/10.1093/nar/gkh121.

[3]   Allen, K.N. and Dunaway-Mariano, D. (2009) Markers of fitness in a successful enzyme superfamily. *Curr. Opin. Struct. Biol.* **19**(6): 658 – 65.
       doi: http://dx.doi.org/10.1016/j.sbi.2009.09.008.

[4]   Burroughs, A.M. *et al.* (2006) Evolutionary genomics of the HAD superfamily: understanding the structural adaptations and catalytic diversity in a superfamily of phosphoesterases and allied enzymes. *J. Mol. Biol.* **361**(5):1003 – 34. doi: http://dx.doi.org/10.1016/j.jmb.2006.06.049.

[5]   Liu, J.Q. *et al.* (1995) Reaction mechanism of L-2-haloacid dehalogenase of Pseudomonas sp. YL. Identification of Asp10 as the active site nucleophile by 18O incorporation experiments. *J. Biol. Chem.* **270**(31):18309 – 12.

[6]   Gross, C., Felsheim, R., and Wackett, L.P. (2008) Genes and enzymes of azetidine-2-carboxylate metabolism: detoxification and assimilation of an antibiotic. *J. Bacteriol.* **190**(14):4859 – 64. doi: http://dx.doi.org/10.1128/JB.02022-07.

[7]   Baker, A.S. *et al.* (1998) Insights into the mechanism of catalysis by the P-C bond-cleaving enzyme phosphonoacetaldehyde hydrolase derived from gene sequence analysis and mutagenesis. *Biochemistry* **37**(26):9305 – 15. doi: http://dx.doi.org/10.1021/bi972677d.

[8]   Morais, M.C. *et al.* (2000) The crystal structure of bacillus cereus phosphonoacet-aldehyde hydrolase: insight into catalysis of phosphorus bond cleavage and catalytic diversification within the HAD enzyme superfamily. *Biochemistry* **39**(34):10385 – 96. doi: http://dx.doi.org/10.1021/bi001171j.

[9]   Collet, J.-F. *et al.* (1997) Human -3-phosphoserine phosphatase: sequence, expression and evidence for a phosphoenzyme intermediate. *FEBS Letters* **408**(3):281 – 284. doi: http://dx.doi.org/10.1016/S0014-5793(97)00438-9.

[10]   Silvaggi, N.R. *et al.* (2006) The X-ray crystal structures of human alpha-phospho-mannomutase 1 reveal the structural basis of congenital disorder of glycosylation type 1a. *J. Biol. Chem.* **281**(21):14918 – 26. doi: http://dx.doi.org/10.1074/jbc.M601505200.

[11]   Koonin, E.V. and Tatusov, R.L. (1994) Computer Analysis of Bacterial Haloacid Dehalogenases Defines a Large Superfamily of Hydrolases with Diverse Specificity: Application of an Iterative Approach to Database Search. *Journal of Molecular Biology* **244**(1):125 – 132. doi: http://dx.doi.org/10.1006/jmbi.1994.1711.

[12]   Dai, J. *et al.* (2009) Analysis of the structural determinants underlying discrimination between substrate and solvent in beta-phosphoglucomutase catalysis. *Biochemistry* **48**(9):1984 – 95. doi: http://dx.doi.org/10.1021/bi801653r.

[13] Lu, Z., Dunaway-Mariano, D., and Allen, K.N. (2008) The catalytic scaffold of the haloalkanoic acid dehalogenase enzyme superfamily acts as a mold for the trigonal bipyramidal transition state. *Proc. Natl. Acad. Sci. U.S.A.* **105**(15):5687 – 92.
doi: http://dx.doi.org/10.1073/pnas.0710800105.

[14] Peisach, E. *et al.* (2004) X-ray crystal structure of the hypothetical phosphotyrosine phosphatase MDP-1 of the haloacid dehalogenase superfamily. *Biochemistry* **43**(40):12770 – 9.
doi: http://dx.doi.org/10.1021/bi0490688.

[15] Galburt, E.A. *et al.* (2002) Structure of a tRNA repair enzyme and molecular biology workhorse: T4 polynucleotide kinase. *Structure* **10**(9):1249 – 60.
doi: http://dx.doi.org/10.1016/S0969-2126(02)00835-3.

[16] Kamenski, T. *et al.* (2004) Structure and mechanism of RNA polymerase II CTD phosphatases. *Mol. Cell* **15**(3):399 – 407.
doi: http://dx.doi.org/10.1016/j.molcel.2004.06.035.

[17] Lahiri, S.D. *et al.* (2004) Analysis of the substrate specificity loop of the HAD superfamily cap domain. *Biochemistry* **43**(10):2812 – 20.
doi: http://dx.doi.org/10.1021/bi0356810.

[18] Lu, Z., Dunaway-Mariano, D. and Allen, K.N. (2005) HAD superfamily phospho-transferase substrate diversification: structure and function analysis of HAD subclass IIB sugar phosphatase BT4131. *Biochemistry* **44**(24):8684 – 96.
doi: http://dx.doi.org/10.1021/bi050009j.

[19] Tremblay, L.W., Dunaway-Mariano, D. and Allen, K.N. (2006) Structure and activity analyses of *Escherichia coli* K-12 NagD provide insight into the evolution of bio-chemical function in the haloalkanoic acid dehalogenase superfamily. *Biochemistry* **45**(4):1183 – 93.
doi: http://dx.doi.org/10.1021/bi051842j.

[20] Zhang, G. *et al.* (2002) Kinetic Evidence for a Substrate-Induced Fit in Phosphono-acetaldehyde Hydrolase Catalysis. *Biochemistry* **41**(45):13370 – 13377.
doi: http://dx.doi.org/10.1021/bi026388n.

[21] Lu, Z. *et al.* (2009) Structure-function analysis of 2-keto-3-deoxy-D-glycero-D-galactononononate-9-phosphate phosphatase defines specificity elements in type C0 haloalkanoate dehalogenase family members. *J. Biol. Chem.* **284**(2):1224 – 33.
doi: http://dx.doi.org/10.1074/jbc.M807056200.

[22] Nguyen, H.H. *et al.* (2010) Structural determinants of substrate recognition in the HAD superfamily member D-glycero-D-manno-heptose-1,7-bisphosphate phospha-tase (GmhB). *Biochemistry* **49**(6):1082 – 92.
doi: http://dx.doi.org/10.1021/bi902019q.

[23]   Kneidinger, B. *et al.* (2002) Biosynthesis pathway of ADP-L-glycero-beta-D-manno-heptose in *Escherichia coli. J. Bacteriol.* **184**(2):363 – 9.
doi: http://dx.doi.org/10.1128/JB.184.2.363-369.2002.

[24]   Kneidinger, B. *et al.* (2001) Biosynthesis of nucleotide-activated D-glycero-D-manno-heptose. *J. Biol. Chem.* **276**(24):20935 – 44.
doi: http://dx.doi.org/10.1074/jbc.M100378200.

[25]   Kosma, P. *et al.* (1995) Glycan structure of a heptose-containing S-layer glycoprotein of *Bacillus thermoaerophilus. Glycobiology* **5**(8):791 – 6.
doi: http://dx.doi.org/10.1093/glycob/5.8.791.

[26]   Messner, P. (1996) Chemical composition and biosynthesis of S-layers, in *Crystalline Bacterial Cell Surface Proteins*, Sleytr, U.B., *et al.*, Editors. R. G. Landes/Academic Press: Austin, TX. p. 35 – 76.
doi: http://dx.doi.org/10.1016/B978-012648470-0/50007-4.

[27]   Schaffer, C. and Messner, P. (2004) Surface-layer glycoproteins: an example for the diversity of bacterial glycosylation with promising impacts on nanobiotechnology. *Glycobiology* **14**(8):31R-42R.
doi: http://dx.doi.org/10.1093/glycob/cwh064.

[28]   Schaffer, C. *et al.* (1999) Complete glycan structure of the S-layer glycoprotein of *Aneurinibacillus thermoaerophilus* GS4 – 97. *Glycobiology* **9**(4):407 – 14.
doi: http://dx.doi.org/10.1093/glycob/9.4.407.

[29]   Valvano, M.A., Messner, P. and Kosma, P. (2002) Novel pathways for biosynthesis of nucleotide-activated glycero-manno-heptose precursors of bacterial glycoproteins and cell surface polysaccharides. *Microbiology* **148**(Pt 7):1979 – 89.

[30]   Wang, L. *et al.* (2010) Divergence of biochemical function in the HAD superfamily: D-glycero-D-manno-heptose-1,7-bisphosphate phosphatase (GmhB). *Biochemistry* **49**(6):1072 – 81.
doi: http://dx.doi.org/10.1021/bi902018y.

Experimental Standard Conditions of Enzyme Characterizations,
September 13th – 16th, 2009, Rüdesheim/Rhein, Germany

79

SUGGESTIONS FOR A PROTEIN SPECIES IDENTIFIER SYSTEM

HARTMUT SCHLÜTER[1,*], HERMANN-GEORG HOLZHÜTTER[2], ROLF APWEILER[3] AND PETER R. JUNGBLUT[4]

[1]Institute of Clinical Chemistry, University Medicine Hamburg-Eppendorf, Martinistr. 53, 20246 Hamburg, Germany

[2]Computational Systems Chemistry, Charité – University Medicine Berlin, Monbijoustr. 2, 10117 Berlin, Germany

[3]EMBL Outstation Hinxton, European Bioinformatics Institute, Wellcome Trust Genome Campus, Hinxton, Cambridge, CB10 1SD, U.K.

[4]Max Planck Institute for Infection Biology, Core Facility Protein Analysis, Charitéplatz 1, 10177 Berlin, Germany

E-Mail: *hschluet@uke.de

Received: 25th January 2010 / Published: 14th September 2010

ABSTRACT

Protein variants, which vary in their exact chemical composition, and which are coded by one gene or by a paralogous or orthologous gene or alleles of that gene, are called protein species. The term protein species covers splicing variants, truncated proteins and post-translational modified proteins, and is defined chemically in contrast to the term isoform, which is defined genetically. The impact of the knowledge of the exact chemical composition of a protein species is determined by the relationship between its composition and its function. Since centuries it is known that post-translational modifications such as phosphorylation critically determine the activity status of enzymes. Proteolytic truncations can activate proteases, peptide hormones or receptors. However, despite of this knowledge, the relationship between the exact chemical composition of a protein and its function is not sufficiently considered in many protein investigations. In many of the current proteomics studies protein identification is based on sequence coverage significantly lower than 100%. Post-translational modifications are more or less ignored. A second drawback concerning the comprehensive description of protein species derives from the absence of an identifier system, which describes their exact chemical composition. Therefore, up to now we have to deal with a huge ambiguity concerning the identity of a protein and its function. In the past, functions were assigned to genes, implicating that the full information for the function is encoded in the DNA sequence. Now it becomes obvious that both different modifications and different combinations result in different protein species with different functions. An identifier system for protein species allows the assignment of a defined function to a defined protein species, which is determined by its exact chemical composition. The protein species identifier system was introduced in 2009 by Schlüter *et al.* and is presented here.

INTRODUCTION

In the past twenty years there was a tremendous increase of knowledge about proteins. The progress in this area was accelerated by the development of new methods in both molecular biology and in chemical structure analysis of proteins. In particular, the development of the soft ionization techniques – ESI (electrospray ionization) [1] and MALDI (matrix assisted laser desorption/ionization) [2] – in mass spectrometry (MS) improved the analysis of the protein composition. Furthermore, the complete sequencing of genomes (see [3,4]) extended the knowledge about proteins. At the same time further questions about the functions of the gene products – the proteins – came up.

From proteomics approaches (see [5]) as well as classical biochemical investigations we know since many years that one gene codes not only for one but for many gene products. These different products can be created by alternative splicing, proteolytic processing of proteins subsequent to the protein synthesis at the ribosome and/or post-translational processing with regard to the covalent addition of functional groups towards residues of the amino acids within the protein. At present, the database UniMod lists more than 600 post-translational modifications (UniMod version in December 2009). This huge quantity of individual post-translational modifications is responsible for a large number of products, the protein species, which can arise from one single gene. The term protein species was introduced by Jungblut *et al.* in 1996 [6]. It was extended for proteomics in 2008 [7], because according to the nomenclature rules of IUBMB the term "protein isoform" does not describe proteins which are encoded by one single gene but proteins with the same function encoded by different genes [8]. Nielsen *et al.* [9] performed a study to evaluate the extent of protein modification. The authors stated that the current estimation of the number of different protein molecules in human beings is close to a million when combining the complexity generated by alternative splicing with that produced by PTMs. This is roughly 50 protein species per gene. How relevant is this huge protein diversity? Since centuries it is well known that covalent modifications, such as phosphorylation, critically determine the activity status of enzymes [10]. The type of linkage of ubiquitins in polyubiquitin chains determines whether a protein is degraded (linkages via lysine 48) or acts as a signal (linkages via lysine 63) within the cell [11]. For proteins, which were investigated in depth often many different functions are listed. For example, it was found out that Hsp70 is involved not only in chaperoning but also in cell growth, apoptosis and genetic recombination [12]. As a result of covalent modification the function of a defined protein can change completely. Another enzyme, GAPDH which is integrated in glycolysis, will induce an apoptosis after being nitrosylated [13]. Besides the covalent modifications of protein side chains, the modification of the amino acid sequence itself is also important for the function. It is well known that enzymes such as thrombin [14] are activated by proteases, which remove one or more peptides from the enzyme by hydrolysis of the peptide bond. A difference in the amino acid sequence of a protein encoded by one single gene can also be induced by alternative splicing: Two splice variants are known from the angiotensin converting enzyme (ACE) gene. The somatic splice variant is involved in blood pressure regulation. In contrast, the ACE splice variant present in testis is responsible for sperm maturation [15].

These examples highlight the importance of the relationship between the chemical composition of a protein and its function. Although the significance of this relationship is not questioned any more, a system which allows an unambiguous assignment of the chemical composition of a protein to its function is still missing. Thus, we propose a protein species identifier system for the description of the chemical composition of every protein species. This system is based on the suggestion published recently [16] and is extended here according to the results of the discussions of the ESCEC meeting in 2009. This suggestion is a framework which has to grow and to be optimized according to the needs of the operators of the protein species identifier system (PSIS). In the near future software will be developed which simplifies the conversion of data about protein species into the PSIS-description.

THE PROTEIN SPECIES IDENTIFIER SYSTEM

The protein species identifier system (PSIS) consists of descriptors that allow the determination of every known aspect of the chemical composition of a protein species.

The description of a defined protein species starts on the level of the coding gene *(level A)* (Table 1). The descriptor consists of the entry gene name according to the UniProt knowledgebase followed by the species name. Both terms are preceded by G, which here stands for gene. In the case of the angiotensin-converting enzyme (ACE), the descriptor for the coding gene is [G_ACE_human]. The second descriptor *(level B)* provides information about nucleotide polymorphisms (NP). The descriptor starts with NP followed by the NP – accession number according to dbSNP of the NCBI (http://www.ncbi.nlm.nih.gov/sites/entrez? db = snp), *e.g.* [NP_rs4331]. On *level C* the initial amino acid sequence of the protein synthesized at the ribosome is described: This descriptor includes the accession number (AC) according to UniProt. For instance, the *level C* descriptor for the angiotensin-converting enzyme is [AC_P12821]. *Level D* refers to splicing. A splicing variant with a deletion (SD) is explained by the localisation of the missing amino acids (*level D-1*). [SD_81 – 97] indicates that in a given protein sequence the amino acids between position 81 and 97 are missing because of splicing. The testis-specific angiotensin-converting enzyme can be described by [SD_1 – 640]. *Level D-2*: The splicing variant containing an insertion (SI), is explained a) by the number of the amino acid, behind which one or several amino acids were inserted. The inserted amino acid(s) are given by the one-letter amino acid code. A *level D-2* descriptor may be: [SI_43_LELFVMFL].

The *level E* descriptor contains information about truncated amino acids (T). For example, active thrombin is generated by the proteolytic removal of the amino acids 1 – 24 (signal peptide), the amino acids 25 – 43 (pro-peptide), the amino acids 44 – 198 (activation peptide fragment 1) and the amino acids 199 – 327 (activation peptide fragment 2). Therefore, the *level E* descriptor for active thrombin is [T_1 – 327].

Post-translational modifications (P) are explained by the *level F* descriptor. The number following P_ indicates the position of the modified amino acid. The second number names the type of the post-translational modification according to the UniMod accession number [17]. In the following example the number 21 indicates a phosphorylation. Seo *et al.* [18] reported that glyceraldehyde-3-phosphate dehydrogenase can be phosphorylated at Thr-75; Ser-122; Ser-148; Thr-229; Thr-237 and Ser-312. For a protein species, which is phosphorylated at all of these positions, the *level F* descriptor is [P_75 – 122 – 148 – 229 – 237 – 312_21].

Cofactors (C) are described by the *level G* descriptor. The cofactors of the human metalloprotease ACE (P12821), for example, are two Zn^{2+}-ions, bound by the amino acids at the positions 390, 394 and 418 as well as 988, 992 and 1016. Therefore, the descriptors of human ACE are [C_ 390 – 394 – 418_Zn] + [C_988 – 992 – 1016_Zn].

If additional descriptors are needed, further levels can be introduced. Every new descriptor should start with a short unambiguous letter describing the aspect of the chemical composition or another important property of a protein in relationship to its chemical composition.

The prefinal descriptor (*level Y*) lists the versions of the data bases (DB) which were used for the identification or description of the exact chemical composition of the protein species. For UniProt retrieved in January 2010 the descriptor is [DB_UniProt_15.12].

The final descriptor (*level Z*) describes the function (F) of an enzyme with the appropriate EC number. The *level Z* descriptor for *e. g.* Glyceraldehyde-3-phosphate dehydrogenase is [F_EC = 1.2.1.12].

Table 1. Listing of the descriptors and their terms of the protein species identifier system (PSIS).

Descriptor-Level	1st Term: Defined aspect of the chemical composition of the protein species	2nd Term: Name or description Recommended data base	3rd Term Further description
	Symbol	*Example*	*Example*
A	Gene	Gene Name UniProt	Species
	G	*ACE*	*human*
B	Nucleotide polymorphisms	Accession number dbSNP (NCBI)	–
	NP	*rs4331*	
C	Initial amino acid sequence of the protein synthesized at the ribosome *AC*	Accession number UniProt *P12821*	–
D-1	Splicing variant	Number of the first and the last amino acid within the sequence which is deleted by splicing	–
	SD	*1 – 640*	
D-2	Splicing variant	Number of amino acid, which precedes the sequence, which was inserted by splicing	Sequence of the inserted peptide
	SI	*43*	*LELFVMFL*
E	Truncated amino acids	Sequence described by the first & the last number of the amino acids within the removed sequence	–
	T	*1 – 29*	
F	Post-translational modifications	Amino acid(s), which are modified	Accession number of the post-translation modification UniMod
	P	*75 – 122 – 148 – 229 – 237 – 312*	*21*
G	Cofactors	Amino acid(s), which bind the cofactor	Symbol describing the cofactor
	C	*390 – 394 – 418*	*Zn*
Y	Data base *DB*	Name of the data base *UniProt*	Version number *15.12*
Z	Function *F*	EC Number *EC = 1.2.1.12*	–

RULES FOR THE PROTEIN SPECIES IDENTIFIER SYSTEM

The general rules for the protein species identifier system (PSIS) are:

1. Descriptors must rely on experimental data, *e. g.* mass spectrometric analysis, immunological methods, knock-out experiments, over-expression experiments.

1.1. At least one descriptor needs to be given, which is either the descriptor for the gene (*level A* descriptor) or the descriptor for the protein (*level C* descriptor) or the function (*level Z* descriptor). Please note that the protein species level is cannot be obtained until information about *level C* and/or level *D* and *E* are given.

1.2. Every descriptor is given in square bracket characters.

1.3. One and the same descriptor can be used several times. For example, if a protein species carries post-translational modifications each post-translational modification is described by its own descriptor. For example the post-translational modification descriptors for a protein, phosphorylated (UniMod accession number: 21) at amino acid number 165 and sulfonated (UniMod accession number: 40) at amino acid number 223, are [P_165_21] + [P_223_40].

1.4. Descriptors are separated by a plus sign.

1.5. Every descriptor is composed by one or more terms.

2.1. A term, for example, is a symbol (*e. g.* NP), an accession number, a number indicating the position of an amino acid or two numbers separated by a hyphen indicating a partial amino acid sequence.

2.2. The underscore character separates terms.

2.3. The first term is a symbol, which is an abbreviation of the individual descriptor explaining a defined aspect of the chemical composition of the protein species, *e. g.* the identity of the gene (symbol: G) coding the protein species.

2.4. The second term refers to *e. g.* the gene name (*level A* descriptor), an accession number (*level B* descriptor and *level C* descriptor) and a partial sequence (*level D-1* or *E* descriptor) described by the first and the last number of the amino acids within the sequence, which is a part of a complete sequence of a defined protein or the position (number within the sequence) of an amino acid (*level D-2, E* or *F* descriptors). If several amino acids within one protein species are modified by the same moiety (*level F* descriptor), or involved in the binding of a cofactor (*level G* descriptor) every number of the concerned amino acids is listed, each number separated by a hyphen.

2.5 The third term refers *e. g.* to the species (*level A* descriptor), to the amino acid sequence of the inserted peptide (*level D-2* descriptor), to the type of post-translational modification (*level F* descriptor) or to a symbol for the cofactor (*level G* descriptor).

2.6 If necessary further terms can be added.

Optional

If the experimental data were obtained by immunological methods (ia = immunological analysis) such as Western Blots the protein accession number (*level C*) must be given. In this case the epitope(s), which is (are) recognized by the antibody, should be described (give the numbers of the amino acids within the epitope), provided the epitopes are known. The term describing an immunological analysis and the epitope of the antibody is for example ia_456 – 464.

REFERENCES

[1]   Fenn, J.B., Mann, M., Meng, C.K., Wong, S.F., Whitehouse, C.M. (1989) Electrospray ionization for mass spectrometry of large biomolecules. *Science* **246**:64 – 71. doi: http://dx.doi.org/10.1126/science.2675315.

[2]   Karas, M., Hillenkamp, F. (1988) Laser desorption ionization of proteins with molecular masses exceeding 10,000 daltons. *Anal. Chem.* **60**:2299 – 2301. doi: http://dx.doi.org/10.1021/ac00171a028.

[3]   Lander E.S., *et al.* (2001) Initial sequencing and analysis of the human genome. *Nature* **409**:860 – 921. doi: http://dx.doi.org/10.1038/35057062.

[4]   Finishing the euchromatic sequence of the human genome. *Nature* (2004) **431**:931 – 945. doi: http://dx.doi.org/10.1038/nature03001.

[5]   Klose, J., Nock, C., Herrmann, M., Stuhler, K., Marcus, K., Bluggel, M., Krause, E., Schalkwyk, L.C., Rastan, S., Brown, S.D., Bussow, K., Himmelbauer, H., Lehrach, H. (2002) Genetic analysis of the mouse brain proteome. *Nat. Genet.* **30**:385 – 393. doi: http://dx.doi.org/10.1038/ng861

[6]   Jungblut, P., Thiede, B., Zimny-Arndt, U., Muller, E.C., Scheler, C., Wittmann-Liebold, B., Otto, A. (1996) Resolution power of two-dimensional electrophoresis and identification of proteins from gels. *Electrophoresis* **17**:839 – 847. doi: http://dx.doi.org/10.1002/elps.1150170505.

[7]   Jungblut, P.R., Holzhütter, H.G., Apweiler, R., Schlüter, H. (2008) The Speciation of the Proteome. *Chem. Cent. J.* **2**:16. doi: http://dx.doi.org/10.1186/1752-153X-2-16

[8]   Joint Commission on Biochemical Nomenclature IUPAC-IUBMB: Nomenclature of multiple forms of enzymes. In: *Biochemical Nomenclature and Related Documents* 2nd edition. Edited by: Liébecq C. Colchester: Portland Press 1992.

[9]   Nielsen, M.L., Savitski, M.M., Zubarev, R.A. (2006) Extent of modifications in human proteome samples and their effect on dynamic range of analysis in shotgun proteomics. *Mol. Cell Proteomics* **5**:2384 – 2391. doi: http://dx.doi.org/10.1074/mcp.M600248-MCP200.

[10]   Riou, J.P., Claus, T.H., Pilkis, S.J. (1978) Stimulation of glucagon of *in vivo* phosphorylation of rat hepatic pyruvate kinase. *J. Biol. Chem.* **253**:656–659.

[11]   Hochstrasser, M. (2009) Origin and function of ubiquitin-like proteins. *Nature* **458**:422–429.
doi: http://dx.doi.org/10.1038/nature07958.

[12]   Morishima, N. (2005) Control of cell fate by Hsp70: more than an evanescent meeting. *J. Biochem.* (Tokyo) **137**:449–453.
doi: http://dx.doi.org/10.1093/jb/mvi057.

[13]   Hara, M.R., Cascio, M.B., Sawa, A. (2006) GAPDH as a sensor of NO stress. *Biochim. Biophys. Acta* **1762**:502–509.

[14]   Lane, D.A., Philippou, H., Huntington, J.A. (2005) Directing thrombin. *Blood* **106**:2605–2612.
doi: http://dx.doi.org/10.1182/blood-2005-04-1710.

[15]   Woodman, Z.L., Schwager, S.L., Redelinghuys, P., Chubb, A.J., van der Merwe, E.L., Ehlers, M.R., Sturrock, E.D. (2006) Homologous substitution of ACE C-domain regions with N-domain sequences: effect on processing, shedding, and catalytic properties. *Biol. Chem.* **387**:1043–1051.
doi: http://dx.doi.org/10.1515/BC.2006.129.

[16]   Schlüter, H., Apweiler, R., Holzhutter, H.G., Jungblut, P.R. (2009) Finding one's way in proteomics: a protein species nomenclature. *Chem. Cent. J.* **3**:11.
doi: http://dx.doi.org/10.1186/1752-153X-3-11.

[17]   http://www.unimod.org/modifications_list.php?:

[18]   Seo, J., Jeong, J., Kim, Y.M., Hwang, N., Paek, E., Lee, K.J. (2008) Strategy for comprehensive identification of post-translational modifications in cellular proteins, including low abundant modifications: application to glyceraldehyde-3-phosphate dehydrogenase. *J. Proteome Res.* **7**:587–602.
doi: http://dx.doi.org/10.1021/pr700657y.

 Beilstein-Institut

Experimental Standard Conditions of Enzyme Characterizations,
September 13th – 16th, 2009, Rüdesheim/Rhein, Germany

87

STANDARD FORMATS FOR PRESENTATION OF SPECTROSCOPIC DATA ON ENZYMES

RICHARD CAMMACK

Pharmaceutical Sciences Research Division, King's College London, 150 Stamford Street, London SE1 9NH, U.K.

E-Mail: richard.cammack@kcl.ac.uk

Received: 10th February 2010 / Published: 14th September 2010

ABSTRACT

Spectroscopic methods are often used to follow the course of enzyme-catalysed reactions. UV/visible spectrophotometry is the most common, but a wide range of other spectroscopic techniques, including infrared and nuclear magnetic resonance, as well as mass spectrometry, are in use. Spectroscopy and spectrometry are also used in the characterization of the enzymes themselves, and in the identification and quantification of substrates and cofactors. Hitherto, there has been no formal requirement to archive original spectroscopic data, as there is for protein structures and gene sequences. However, the funding agencies increasingly expect grantees to have policies on data sharing, and to deposit all types of experimental data. Spectroscopic data are now conveniently acquired in digital form, but apart from printed documents, there are no universally accepted formats for data storage. Spectra are produced by proprietary software written by instrument manufacturers to run their own instruments. Data formats are nonstandard and may be difficult to read directly. Standard, vendor-neutral data formats have been established for certain types of spectroscopy, such as JCAMP-DX (from the Joint Committee on Atomic and Molecular Physical Data eXchange) for several types of spectroscopy, including infrared and NMR. The details of each format necessarily depend on the type of spectroscopy. There are parallel developments of criteria for meta-data and data validation. Standard formats will facilitate the use of electronic notebooks. The extension of these formats to different types of spectroscopy and spectrometry will facilitate their linkage to other chemical information such as molecular structure. ASCII formats such as JCAMP-DX or, more recently, XML formats such as CML (Chemical Markup Language) satisfy the data-storage requirements. They are readable by generic, open-source software. The routine deposition of spectra in electronic repositories (databanks) will benefit the biochemical community by making them available for further analysis and data mining.

INTRODUCTION

Many enzyme assays employ spectroscopy as a non-destructive method to measure kinetics of enzyme-catalysed reactions. UV/visible spectrophotometry is commonly used for continuous measurement of substrates or products. Fluorimetry offers higher sensitivity if fluorogenic substrates are available. For such purposes, the output is a two-dimensional graph of the concentration of a particular species as a function of time, from which reaction rates can be calculated. Often all that is required is a single absorbance reading at each time point. However modern spectrophotometers make it easy to measure the whole spectrum, which allows further operations such as spectral deconvolution and baseline subtraction to extract the concentration data from a sample containing multiple chromophores.

On early spectrophotometers, spectroscopic data were recorded on paper charts. The spectra were then transcribed into figures in paper publications. Operators became adept at recognizing the shapes and details of such spectra, but further analysis was very limited before spectrometers interfaced to computers were the norm. For publication of the data, the spectra were recorded at low resolution; they were often redrawn by hand, so that details were lost. The paper charts could be digitized, but this is laborious and entails loss of resolution. Meanwhile, the electronic versions of the spectra recorded on diverse computer systems soon were rendered unreadable by the rapid obsolescence of computer operating systems and data storage media. As a result, the data from many careful studies, sometimes on material that is no longer accessible, is now unavailable for analysis and comparison.

In the characterization of the enzymes and their cofactors, a wider range of spectroscopic methods is used [1], including NMR spectroscopy, Fourier-transform infrared (FTIR), circular dichroism (CD) and mass spectrometry. More specialist techniques can provide additional information about enzyme mechanisms, such as electron paramagnetic resonance (EPR) for flavins and transition-metal ions. Spectroscopic data are rich in information, and may be analyzed in different ways to extract information about enzymes and the reactions they catalyse, including:

- observation of transient intermediates in the enzyme-catalysed reaction;
- quantitative analysis, by comparison with spectra of standard samples;

- resolution of the spectra into their principal components, for example Gaussian line shapes;

- extraction of fundamental parameters of the enzyme-bound species, for example by simulation of the spectra using appropriate theory.

THE IMPORTANCE OF RETAINING ORIGINAL SPECTROSCOPIC DATA ON ENZYMES

During a typical study, many spectra are recorded. Techniques such as Fourier-transform NMR and pulsed EMR generate extensive sets of multidimensional data. Because of limitations of space, careful selection is required of data for publication, so the results are usually presented in the form of derived parameters. A whole spectrum may be reduced to a single data point in a two-dimensional plot. This was inevitable in the past, when the storage of large quantities of data was difficult and costly. Now that virtually infinite electronic storage capacity is available, the raw data can be preserved. Often acquired with much effort and expense, these are an extremely valuable resource for further analysis by different approaches. They are required for comparison with other experimental data, and further theoretical analysis.

From the point of view of the STRENDA initiative, it is important that when enzyme data that rely on spectroscopic techniques are published, full details of the experiment and results are provided. Ideally all the original spectral data should be made available, not just the derived data required to create the published figures [2, 3]. Journals, which require deposition of gene sequences or protein structures in open-access databases, do not mandate this for key spectroscopic data. They could not print all the experimental data from a study; for example, results that are valid but lead to negative conclusions are rarely published. In chemistry, it has been estimated that more than 99% of spectra that are recorded are lost [4]. Many data languish on computer disks "trapped by technical, legal and cultural barriers – a problem that open-data advocates are only just beginning to solve" [5]. However this situation is likely to change. Whereas in the past there has been little incentive to share such information, it is now becoming mandatory for publicly funded research to present all the experimental data. Many funding bodies now have policies that require data sharing.

This review considers the issues that have to be addressed for acquisition and long-term storage of spectroscopic data, with particular reference to those on enzymes. Seamless data transfer will be facilitated if spectrometers can output their data in standard formats; if the experimental data are recorded in electronic notebooks; and if repositories to store and retrieve the data are readily accessible on the internet. However the lack of data standards inhibits the full exploitation of spectroscopic data.

DIGITAL DATA FORMATS

Typically, the output from a spectrometer is captured as $x–y$ data, such as a curve of absorbance *vs.* wavelength or frequency, as in UV/visible or FTIR spectrophotometry. In digital format this is stored in a series of data points representing a two-dimensional plot. Alternatively the output may take the form of a list of peak positions and amplitudes, as in NMR or mass spectrometry. Additional dimensions may also be introduced by varying other parameters, such as time or temperature, yielding larger data sets (n-tuple arrays). Other types of measurement, such as chromatography, also produce x-y plots. These data can be captured electronically and conveniently stored in computer databases and retrieved for further analysis. From these data, parameters are extracted for publication. Printed documents are a way to save the data in a permanent and easily readable format. However, if publications are the only source of spectroscopic results, most of the original information in the spectra is lost [2]. It is difficult to store and retrieve the original data, and in particular to search the spectra for particular features.

Digital data formats for spectroscopy were introduced by manufacturers of spectroscopic instruments, when these were interfaced to computers. This is very convenient; the instrumental parameters are stored automatically with the spectral data, and can be stored and exchanged between users of the same instrument. However incompatibility between the bespoke systems was an issue. When computer memory was at a premium, formats were highly compressed, such as binary data, and were not human readable. Moreover they were proprietary, needing expensive software to read them. The software written for one spectrometer is usually incompatible with data from other manufacturers' instruments, and even older instruments from the same manufacturer. This requires special programs to be written for inter-conversion of formats. Such software tends to become obsolete in time, as computer operating systems evolve.

DATA STORAGE

After a time, retrieval of spectroscopic data raises other issues. The files need to be systematically documented. Whereas an expert human reader can readily recognize the subtle similarities and differences in shapes of different spectra, it is very difficult to program a computer to recognize the salient features automatically (famously, it is very difficult by means of computer software to distinguish a picture of a cat from a dog, the basis of the CAPTCHA security system [6]). Moreover, little information can be extracted from free-text fields. Therefore, meta-data are essential for the deposition and retrieval of the spectra. Moreover, there must be sufficient information for the work to be reproduced. For this reason, there has to be a formal system for deposition of information such as the details of sample, aims of the experiment, ownership of the data and credits for the work. This has to be entered by the operator, ideally at the time of the measurement.

Standard file formats

Sharing of spectroscopic data acquired on different instruments requires the introduction of a "lingua franca" or a standard, preferably non-proprietary, format. Otherwise it is necessary, for each manufacturer's spectroscopic format, to have conversion software to and from other formats. In some areas of information technology such standard formats are well established, such as the portable document format (PDF) for printed documents, and the JPG and PNG formats for digital images. When such standards are established, files in these formats can be recognized by web browsers and other generic programs, which help to overcome the problem of obsolescence of specialist software.

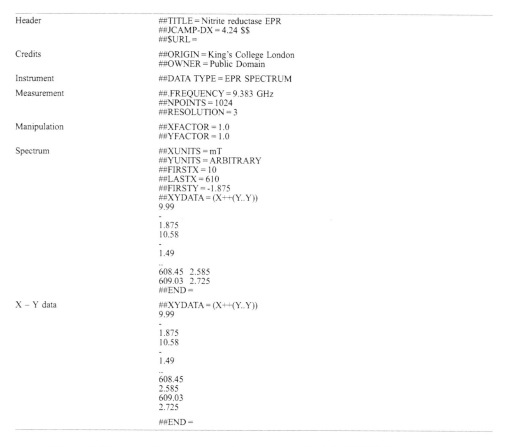

Figure 1. Example of a spectrum, with the layout of the JCAMP-DX file.
The table shows some minimal information for presenting an EPR spectrum.

A number of different standard formats for spectroscopic data have been proposed over the years, with varying levels of acceptance by the community. JCAMP-DX is a flexible format, introduced in the 1980's by the Joint Committee on Atomic and Molecular Physical Data.

It has been used extensively, notably for FTIR [7, 8], NMR [9] and mass spectrometry [10]. It consists of a single file, representing data arrays of 2, 3 or more dimensions. A file comprises labelled data records, each consisting of a flagged data label (usually fairly self-explanatory), and an associated data-set. The file may contain a spectrum or a block of spectra. It also incorporates meta-data to describe the sample(s), measurement conditions and other experimental details. The use of consistent data-labels makes it possible to use generic software to output from several different types of spectroscopy and spectrometry. An example, including some sections of a file for EPR, is shown in Table 1. Some spectrometer manufacturers, notably in FTIR and NMR, offer conversion software to produce spectra in JCAMP-DX format. The format is open-source, and can be read by any program that can handle ASCII text, though special software is needed to view and interpret the spectra. JSpecView is an Open Source applet for viewing JCAMP-DX and AnIML spectra files, allowing zooming and integration [11]. For studies of enzymes, JCAMP-DX protocols of interest in studies of enzymes are for infrared spectroscopy [7] (which can be extended to UV/visible spectrophotometry), NMR spectroscopy [9], electron magnetic resonance (EPR and ESR) spectroscopy [12] and mass spectrometry [10]

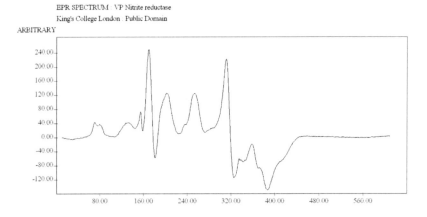

Figure 2. Spectrum plotted from the JCAMP-DX file in Figure 1. The table shows some minimal information for presenting a spectrum. It was plotted using the CHIME plug-in from Symyx [22] (note that this is no longer supported). Sample courtesy of Dr V.O. Popov, Bakh Institute of Biochemistry, Moscow.

Repositories for Spectroscopic data

Specialist technique-specific databanks are being developed for the purpose of collecting and disseminating spectroscopic data. Some of these databases are small enough that they can be downloaded from the internet. BioMagResBank – Biological Magnetic Resonance

Data Bank (BMRB: www.bmrb.wisc.edu) [13] – collects data on NMR spectroscopy of biological entities. It uses NMR-STAR format for proteins [14]. The Protein Circular Dichroism Data Bank (pcddb.cryst.bbk.ac.uk) is a newly-established repository for circular dichroism spectroscopy [15]. Ultraviolet CD is used for characterization of protein structure. It is used to estimate the proportion of secondary structure elements, such as alpha-helix and beta sheet. The PCCDB site is intended to offer an archive of user-deposited CD spectra; this data can be used in conjunction with analyses algorithms [16] to determine the secondary structures of proteins. It includes conversion software to convert spectra recorded in ten different formats on spectrometers from five manufacturers plus synchrotron CD spectra, into an in-house format that is downloadable in a generic text format.

XML formats

There are initiatives to replace the flat-file formats like JCAMP-DX by web-based markup languages. XML has been adopted in many areas of computer technology, including communications and distributed computing. There are XML schemas for various aspects of chemistry and biochemistry. The schemas incorporate data dictionaries to ensure consistent terminology, and validation criteria to ensure data quality. Chemical markup language, CML, is used to represent chemical data and documents [17, 18]. Within CML, an XML vocabulary, CMLSpect, has been developed for spectral data [2]. For analytical chemistry, AnIML (Analytical Information Markup Language), first proposed in 2003, is under development as an IUPAC/ASTM unified standard for spectroscopic and related information [19, 20]. The five initial AnIML techniques are UV/visible and infrared spectrophotometry, mass spectrometry, 1D NMR and chromatography [21].

Once acquired, spectroscopic data may be stored in the institution where the work was done; with the publisher where the work was published; or in a central repository. In each case the repository should have facilities for data deposition, validation, viewing, search and retrieval. The SPECTRa (Submission, Preservation and Exposure of Chemistry Teaching and Research Data) Project, which uses markup languages [4] has pioneered the archiving of primary chemistry data, including crystallographic structures, spectroscopy (principally NMR) and computational chemistry. This initiative has identified a number of features that are necessary for such repositories. Each component has a unique and persistent identifier. There is an embargo system, for research data prior to publication. In addition to the metadata requirements already mentioned, there should be a graphical user interface (GUI) for browsing, navigation, and text-based searches of the text and metadata.

CONCLUSIONS

Deposition of spectroscopic data requires open-source, generic formats. These should be easy to read, archive and retrieve. Repositories must be able to accept multiple instrumental formats, including legacy data from older instruments. Software is required for archiving, retrieval, display and manipulation of spectroscopic data. Meta-data are essential for archiving and retrieval of data, and data-mining. To achieve this will require the expansion of deposition requirements for both published and unpublished data, and funding for software and database development.

ACKNOWLEDGEMENTS

I am greatly indebted to Robert Lancashire (University of the West Indies); Bonnie Wallace (Birkbeck College London) and Bob Janes (Queen Mary University London) and Peter Murray-Rust (University of Cambridge) for discussions and comment, and Dr Vladimir Popov for the data used in Figure 1.

ABBREVIATIONS

ASTM	American Society for Testing and Materials
BioMagResBank	Biological Magnetic Resonance Data Bank
CAPTCHA	Completely Automatic Public Turing Test to Tell Computers and Humans Apart
CD	circular dichroism
CML	chemical markup language
EMR	electron magnetic resonance
EPR	electron paramagnetic resonance
ESR	electron spin resonance
FTIR	Fourier-transform infrared spectroscopy
GUI	graphical user interface
IUPAC	International Union for Pure and Applied Chemistry
JCAMP-DX	Joint committee on atomic and molecular physical data exchange
MDL	MDL Molecular Design Ltd
NMR	nuclear magnetic resonance
PCDDB	Protein Circular Dichroism Data Bank
SPECTRa	Submission, Preservation and Exposure of Chemistry Teaching and Research Data
STRENDA	Standards for Reporting Enzymology Data

REFERENCES

[1]   Reymond, J. L. (2006) Enzyme assays: High-throughput screening, Genetic selection and Fingerprinting, Wiley-VCH, Weinheim.

[2]   Kuhn, S., Helmus, T., Lancashire, R. J., Murray-Rust, P., Rzepa, H. S., Steinbeck, C., and Willighagen, E. L. (2007) Chemical markup, XML, and the world wide web. 7. CMLSpect, an XML vocabulary for spectral data. *Journal of Chemical Information and Modeling* **47**:2015 – 2034.
       doi: http://dx.doi.org/10.1021/ci600531a.

[3]   Cammack, R. (2010) EPR Spectra of Transition-Metal Proteins: the Benefits of Data Deposition in Standard Formats. *Applied Magnetic Resonance* **37**:257 – 266.
       doi: http://dx.doi.org/10.1007/s00723-009-0095-2.

[4]   Downing, J., Murray-Rust, P., Tonge, A. P., Morgan, P., Rzepa, H. S., Cotterill, F., Day, N., and Harvey, M. J. (2008) SPECTRa: The deposition and validation of primary chemistry research data in digital repositories. *Journal of Chemical Information and Modeling* **48**:1571 – 1581.
       doi: http://dx.doi.org/10.1021/ci7004737.

[5]   Nelson, B. (2009) Empty archives. *Nature* **461**:160 – 163.
       doi: http://dx.doi.org/10.1038/461160a.

[6]   Elson, J., Douceur, J. R., Howell, J., and Saul, J. (2007) Asirra: A CAPTCHA that Exploits Interest-Aligned Manual Image Categorization. In *Proceedings of the 14th ACM Conference on Computer and Communication Security* (DiVimercati, S. D. C., Syverson, P., and Evans, D., eds) pp. 366 – 374, Alexandria, VA.

[7]   McDonald, R. S., and Wilks, P. A. (1988) JCAMP-DX – a standard form for exchange of infrared-spectra in computer readable form. *Appl. Spectrosc.* **42**:151 – 162.
       doi: http://dx.doi.org/10.1366/0003702884428734.

[8]   Grasselli, J. G. (1991) JCAMP-DX, a standard format for exchange of infrared-spectra in computer readable form. *Pure Appl. Chem.* **63**:1781 – 1792.
       doi: http://dx.doi.org/10.1351/pac199163121781.

[9]   Davies, A. N., and Lampen, P. (1993) JCAMP-DX for NMR. *Appl. Spectrosc.* **47**:1093 – 1099.
       doi: http://dx.doi.org/10.1366/0003702934067874.

[10]   Lampen, P., Hillig, H., Davies, A. N., and Linscheid, M. (1994) JCAMP-DX for mass-spectrometry. *Appl. Spectrosc.* **48**:1545 – 1552.
       doi: http://dx.doi.org/10.1366/0003702944027840.

[11]   Lancashire, R. J. (2007) The JSpecView project: an Open Source Java viewer and converter for JCAMP-DX, and XML spectral data files. *Chemistry Central Journal* **1**:31.
       doi: http://dx.doi.org/10.1186/1752-153X-1-31.

[12]   Cammack, R., Fann, Y., Lancashire, R. J., Maher, J. P., McIntyre, P. S., and Morse, R. (2006) JCAMP-DX for electron magnetic resonance(EMR). *Pure Appl. Chem.* **78**:613 – 631.
       doi: http://dx.doi.org/10.1351/pac200678030613.

[13]   Ulrich, E. L., Akutsu, H., Doreleijers, J. F., Harano, Y., Ioannidis, Y. E., Lin, J., Livny, M., Mading, S., Maziuk, D., Miller, Z., Nakatani, E., Schulte, C. F., Tolmie, D. E., Wenger, R. K., Yao, H. Y., and Markley, J. L. (2008) BioMagResBank. *Nucleic Acids Res.* **36**:D 402-D 408.
       doi: http://dx.doi.org/10.1093/nar/gkm957.

[14]   Doreleijers, J. F., Nederveen, A. J., Vranken, W., Lin, J. D., Bonvin, A., Kaptein, R., Markley, J. L., and Ulrich, E. L. (2005) BioMagResBank databases DOCR and FRED containing converted and filtered sets of experimental NMR restraints and coordinates from over 500 protein PDB structures. *J. Biomol. NMR* **32**:1 – 12.
       doi: http://dx.doi.org/10.1007/s10858-005-2195-0.

[15]   Wallace, B. A., Whitmore, L., and Janes, R. W. (2006) The Protein Circular Dichroism Data Bank (PCDDB): A bioinformatics and spectroscopic resource. *Proteins-Structure Function and Bioinformatics* **62**:1 – 3.
       doi: http://dx.doi.org/10.1002/prot.20676.

[16]   Whitmore, L., and Wallace, B. A. (2008) Protein secondary structure analyses from circular dichroism spectroscopy: Methods and reference databases. *Biopolymers* **89**:392 – 400.
       doi: http://dx.doi.org/10.1002/bip.20853.

[17]   Murray-Rust, P., and Rzepa, H. S. (2003) Chemical Markup, XML, and the World Wide Web. 4. CML Schema. *J. Chem. Inf. Comput. Sci.* **43**:757 – 772.

[18]   Murray-Rust, P., Rzepa, H. S., and Wright, M. (2001) Development of chemical markup language (CML) as a system for handling complex chemical content. *New J. Chem.* **25**:618 – 634.
       doi: http://dx.doi.org/10.1039/b008780g.

[19]   Julian, R. K. (2003) The IUPAC/ASTM Unified Standard for Analytical Data: AnIML. *Scientific Computing* www.scientificcomputing.com/the-iupac-astm-unified-standard.aspx.

[20]   Davies, A. N. (2007) Herding AnIMLs (no, it's not a spelling mistake): Update on the IUPAC and ASTM Collaboration on Analytical Data Standards. *Chemistry International* **29**(6).

[21]   Lancashire, R. J., and Davies, A. N. (2006) Spectroscopic Data: The Quest for a Universal Format. *Chemistry International* **28**(1).

[22]   Lancashire, R. J. (2000) The use of the Internet for teaching Chemistry. *Anal. Chim. Acta* **420**:239 – 244.
        doi: http://dx.doi.org/10.1016/S0003-2670(00)00895-3.

STRUCTURAL CORRELATES OF PROTEIN MELTING TEMPERATURE

ERIC A. FRANZOSA[1], KEVIN J. LYNAGH[2], AND YU XIA[1,*]

[1]Bioinformatics Program, Dept. of Chemistry, Dept. of Biomedical Engineering, Boston University, 24 Cummington Street, Boston, MA 02215, U.S.A.

[2]Reed College, MS 880, 3203 SE Woodstock Boulevard, Portland, Oregon 97202 – 8199, U.S.A.

E-Mail: *yuxia@bu.edu

Received: 7th January 2010 / Published: 14th September 2010

ABSTRACT

The stability of a protein's native state has important implications for its folding dynamics, function, and evolution. Here we report on a study investigating general relationships between sequence- and structure-based properties of a protein and its empirically determined stability (as measured by melting temperature experiments). Surprisingly, we find that contact density – a sequence-independent measure of protein compactness – is not significantly correlated with protein melting temperature; this property has been previously implicated as a correlate of protein stability in theoretical and evolutionary analyses. After incorporating residue type in the definition of residue-residue contacts, we find that increasing the fraction of hydrophobic contacts in a protein tends to raise melting temperature, consistent with a stabilizing effect, while increasing the fraction of repulsive charge contacts results in a marginally significant decrease in melting temperature, consistent with a destabilizing effect. Our work demonstrates that subtle sequence variation may be an important factor in fine-tuning the stability of a protein fold.

INTRODUCTION

A protein's stability can be thought of as its thermodynamic "preference" for achieving and maintaining the native (folded) state. This evolved property is critical for protein function and for preventing the accumulation of cytotoxic unfolded or misfolded protein forms. Stability can vary among proteins from a single species and among homologues of a given protein across species. A well-known example of the latter case is *Taq* polymerase: a bacterial DNA polymerase evolved for the high temperature environment of its thermophilic source organism, and also a key component in the development of the polymerase chain reaction (PCR).

A popular laboratory measure of protein stability is *melting temperature*. In melting temperature experiments, a solution of a protein in its native state is heated until the point of complete denaturation. The temperature at which the native and denatured states are equally populated at equilibrium is called the melting temperature, and it is inferred from a change in the optical properties of the solution. Melting temperature experiments are frequently carried out on wild type and mutant proteins to investigate the contributions of specific residues to protein stability. The ProTherm database contains a record of many such experiments [1].

Like all protein properties and functions, stability "information" must be encoded by the protein's structure, which is in turn specified by the protein's primary sequence. In principle it should therefore be possible to predict protein stability from sequence and structure data. Along these lines, many methods have been proposed for predicting *changes* in protein stability upon introduction of a mutation; see [2 – 7] for some recent examples. In this work, we aim to understand better which protein properties contribute to variation in experimental protein melting temperatures. Previous studies in this area have generally relied upon binning proteins to look for systematic differences between sequences and structures derived from organisms adapted to low, moderate, and high temperatures; see [8 – 14] for some examples. These studies revealed general and position-specific preferences for certain amino acids and amino acid pairs in different temperature regimes. Our approach has an advantage in that, instead of using "source organism" as a discrete proxy for protein temperature tolerance, we consider a direct, continuous measure of protein stability based on laboratory experiments. We hypothesize that uncovering the protein properties most correlated with melting temperature may provide new insight into the mechanisms underlying protein stability.

METHODS & RESULTS

We collected data for 72 single-domain, wild-type proteins with (i) known structures in the Protein Data Bank [15] and (ii) melting temperature data from at least one experiment reported in the ProTherm database [1]. We considered average melting temperature if more than one experimental value was given. Total surface area and volume were computed for each protein structure using the program MSMS [16]. An *L-by-L contact matrix* was also constructed for each protein structure, where *L* is the protein's length. The (i, j) entry of such a matrix is 1 if the alpha carbons of residues i and j are separated by no more than 7 angstroms (representing contact) and 0 otherwise (representing no contact).

melting temperature (°C)	0.172	-0.067	-0.070	-0.191	-0.116	0.013	0.002	0.027	0.055
	hydrophobic contacts (%)	-0.632	-0.143	-0.338	-0.897	0.060	0.248	0.238	0.217
		polar contacts (%)	-0.539	-0.273	0.430	0.155	0.018	-0.032	0.025
			salt bridges (%)	0.539	0.118	-0.225	-0.240	-0.145	-0.235
				repulsive contacts (%)	0.180	-0.242	-0.262	-0.214	-0.243
					mixed contacts (%)	-0.058	-0.233	-0.225	-0.198
						contact density	0.527	0.450	0.531
							length	0.950	0.962
								surface area (Å²)	0.955
									volume (Å²)

Figure 1. Relationships among protein melting temperature and protein structural properties. For each pair of properties we report the Pearson correlation coefficient (above the diagonal) and a scatter plot (below the diagonal) comparing data for the 72 proteins in our study.

The *contact density* of a protein is the average number of contacts per residue, and serves as a measure of protein "compactness". We also considered the maximum Eigen value of the contact matrix as a more sophisticated notion of contact density [17]; this measure generalizes a residue's degree of connectedness beyond its immediate neighbours. To incorporate sequence data, we then categorized each individual residue-residue contact as one of the following: hydrophobic (both residues are hydrophobic), polar (both residues are polar), mixed (one residue is polar, the other is hydrophobic), salt bridge (residues have opposite charge at physiological pH), and repulsive (residues have the same charge at physiological pH). Correlations reported in the paper and the figure represent Pearson's correlation; one-tailed *p*-values were determined from 10,000 rounds of randomizing permutation. The Figure 1 also contains a scatter plot comparison of each property pair.

Basic protein geometric features (length, surface area, and volume) correlate extremely weakly with melting temperature. The correlation between melting temperature and contact density is also non-significant ($r = 0.013$; $p = 0.475$). This second fact is surprising given the theoretical importance of contact density: in simulated proteins, contact density has been demonstrated as an important determinant of fold designability [17], which is itself proposed to positively correlate with protein stability [18]. These ideas are supported by the observation that (across multiple species) contact density correlates positively with evolutionary rate [19, 20] – an observation that makes sense if proteins of high contact density are very stable and therefore robust against mutation. However, the observed lack of significant correlation between contact density and melting temperature suggests either that contact density is not directly connected to protein stability, or that the connection is too weak to detect using our small dataset (72 proteins). We also investigated a more sophisticated notion of contact density (the maximum Eigen value of the contact matrix), and reached similar conclusions.

One weakness of traditional contact density measurements is that they do not consider protein primary sequence information. Two proteins with the same backbone geometry will have the same contact density, even if their amino acid sequences are very different. To test the importance of amino acid sequence in determining melting temperature, we classified contacts in our protein dataset based on the biochemical nature of the involved pairs of amino acids. We identified a slight tendency for proteins with a larger fraction of hydrophobic contacts to have elevated melting temperature ($r = 0.172$; $p = 0.072$). Note that, although larger proteins tend to contain a larger fraction of hydrophobic contacts, the poor correlation between melting temperature and properties such as length, surface area, and volume precludes explanation based on a size effect. Maximizing hydrophobic contacts is a critical driving force in protein folding [21], and so it is reasonable to speculate that a larger fraction of such contacts might further stabilize a protein fold. Surprisingly, the correlation between melting temperature and fraction of salt bridges (contacts between oppositely charged residues) was weakly negative and non-significant ($r = -0.070$; $p = 0.275$). This seemingly counter-intuitive observation that salt bridge interactions do not contribute positively to protein stability can be explained by the large unfavourable free energy cost of desolvating these charged residues in the first place [22]. Finally, the fraction of repulsive contacts (interactions between like charged residues) showed a marginally significant, negative correlation with melting temperature ($r = -0.191$; $p = 0.053$), consistent with a destabilizing effect.

DISCUSSION

Although the significance of contact density in protein evolution and design has been theoretically and empirically demonstrated, it does not seem to play a significant role in determining protein melting temperature in our dataset. This observation can be rationalized by returning to the *Taq* polymerase example from the Introduction. *Taq* polymerase has a higher melting temperature than its counterpart in *Escherichia coli*. However, from a structural perspective, the polymerase domains of the two proteins are almost identical [23] and hence their computed contact densities will also be very similar. The same degree of similarity is not found at the sequence level. Indeed, a BLAST search reveals that the polymerase domains of the two proteins have fewer than 50% identically aligned residues [24]. This fraction is sufficient to produce highly similar folds as a result of the many-to-one nature of the sequence-structure relationship. At the same time, these numerous differences allow for great variation in the types of residue-residue contacts found in the two protein backbones, which may play a role in explaining the observed difference in their melting temperatures. We show here that a decrease in the fraction of repulsive contacts in a protein results is a marginally significant increase in melting temperature. An increase in the fraction of hydrophobic contacts was also associated with melting temperature increase. These findings provide support for the role of amino acid sequence in fine-tuning the stability of a protein fold.

The greatest limitation we have encountered thus far is the low availability of wild-type proteins of known structure and melting temperature. In the future we aim to expand our dataset through the addition of multi-domain proteins or by structural homology modelling. We are further limited by the general noisiness of protein melting temperature data and the fact that these data are derived from many independent experiments. Finally, the contact-based parameters presented here were based on alpha-carbon models of protein structures (by our own approximation). We will next expand the work to consider all-atom models, which may provide a more realistic picture of the contacts between residues. Nevertheless, our study here demonstrates that bioinformatics analysis of protein stability data can provide insights into the structural determinants of protein thermodynamics.

ACKNOWLEDGMENTS

EAF was supported by an IGERT Fellowship through NSF grant DGE-0654108 awarded to the BU Bioinformatics Program. KJL was supported by an REU Fellowship through NSF grant CHE-0649114. YX was supported by a Research Starter Grant in Informatics from the PhRMA Foundation.

REFERENCES

[1]   Kumar, M.D., *et al.*(2006) ProTherm and ProNIT: thermodynamic databases for proteins and protein-nucleic acid interactions. *Nucleic Acids Res.* **34**(Database issue): D204–6.
doi: http://dx.doi.org/10.1093/nar/gkj103.

[2]   Capriotti, E. *et al.* (2008) A three-state prediction of single point mutations on protein stability changes. *BMC Bioinformatics* **9**(Suppl 2):S6.
doi: http://dx.doi.org/10.1186/1471-2105-9-S2-S6.

[3]   Cheng, J., Randall, A. and Baldi, P. (2006) Prediction of protein stability changes for single-site mutations using support vector machines. *Proteins* **62**(4):1125–32.
doi: http://dx.doi.org/10.1002/prot.20810.

[4]   Gromiha, M.M. (2007) Prediction of protein stability upon point mutations. *Biochem. Soc. Trans.* **35**(Pt 6):1569–73.
doi: http://dx.doi.org/10.1042/BST0351569.

[5]   Huang, L.T. *et al.* (2007) Prediction of protein mutant stability using classification and regression tool. *Biophys. Chem.* **125**(2–3):462–70.
doi: http://dx.doi.org/10.1016/j.bpc.2006.10.009.

[6]   Masso, M. and Vaisman, II (2008) Accurate prediction of stability changes in protein mutants by combining machine learning with structure based computational mutagenesis. *Bioinformatics* **24**(18):2002–9.
doi: http://dx.doi.org/10.1093/bioinformatics/btn353.

[7]   Parthiban, V. *et al.* (2007) Structural analysis and prediction of protein mutant stability using distance and torsion potentials: role of secondary structure and solvent accessibility. *Proteins* **66**(1):41–52.
doi: http://dx.doi.org/10.1002/prot.21115.

[8]   Cambillau, C. and Claverie, J.M. (2000) Structural and genomic correlates of hyperthermostability. *J. Biol. Chem.* **275**(42):32383–6.
doi: http://dx.doi.org/10.1074/jbc.C000497200.

[9]   Gianese, G., Bossa, F., and Pascarella, S. (2002) Comparative structural analysis of psychrophilic and meso- and thermophilic enzymes. *Proteins* **47**(2):236–49.
doi: http://dx.doi.org/10.1002/prot.10084.

[10]   Gromiha, M.M. (2001) Important inter-residue contacts for enhancing the thermal stability of thermophilic proteins. *Biophys. Chem.* **91**(1): 71–7.
doi: http://dx.doi.org/10.1016/S0301-4622(01)00154-5.

[11]   Kannan, N. and Vishveshwara, S. (2000), Aromatic clusters: a determinant of thermal stability of thermophilic proteins. *Protein Eng.* **13**(11):753 – 61.
doi: http://dx.doi.org/10.1093/protein/13.11.753.

[12]   Kumar, S., Tsai, C.J., and Nussinov, R. (2001) Thermodynamic differences among homologous thermophilic and mesophilic proteins. *Biochemistry* **40**(47):14152 – 65.
doi: http://dx.doi.org/10.1021/bi0106383.

[13]   Pack, S.P. and Yoo, Y.J. (2004) Protein thermostability: structure-based difference of amino acid between thermophilic and mesophilic proteins. *J. Biotechnol.* **111**(3):269 – 77.
doi: http://dx.doi.org/10.1016/j.jbiotec.2004.01.018.

[14]   Szilagyi, A. and Zavodszky, P. (2000) Structural differences between mesophilic, moderately thermophilic and extremely thermophilic protein subunits: results of a comprehensive survey. *Structure* **8**(5):493 – 504.
doi: http://dx.doi.org/10.1016/S0969-2126(00)00133-7.

[15]   Berman, H.M. *et al.* (2000) The Protein Data Bank. *Nucleic Acids Res.* **28**(1):235 – 42.
doi: http://dx.doi.org/10.1093/nar/28.1.235.

[16]   Sanner, M.F., Olson, A.J., and Spehner, J.C. (1996) Reduced surface: an efficient way to compute molecular surfaces. *Biopolymers* **38**(3):305 – 20.
doi: http://dx.doi.org/10.1002/(SICI)1097-0282(199603)38:3<305::AID-BIP4>3.0.CO;2-Y.

[17]   England, J.L. and Shakhnovich, E.I. (2003) Structural determinant of protein designability. *Phys. Rev. Lett.* **90**(21):218101.
doi: http://dx.doi.org/10.1103/PhysRevLett.90.218101.

[18]   Bloom, J.D. *et al.* (2006) Protein stability promotes evolvability. *Proc. Natl. Acad. Sci. U.S.A.* **103**(15):5869 – 74.
doi: http://dx.doi.org/10.1073/pnas.0510098103.

[19]   Zhou, T., Drummond, D.A. and Wilke, C.O. (2008) Contact density affects protein evolutionary rate from bacteria to animals. *J. Mol. Evol.* **66**(4):395 – 404.
doi: http://dx.doi.org/10.1007/s00239-008-9094-4.

[20]   Bloom, J.D. *et al.* (2006) Structural determinants of the rate of protein evolution in yeast. *Mol. Biol. Evol.* **23**(9):1751 – 61.
doi: http://dx.doi.org/10.1093/molbev/msl040.

[21]   Murphy, K.P. (2001) Stabilization of protein structure. *Methods Mol. Biol.* **168**:1 – 16.

[22]   Hendsch, Z.S. and Tidor, B. (1994) Do salt bridges stabilize proteins? A continuum electrostatic analysis. *Protein Sci.* **3**(2):211 – 26.

[23]   Eom, S.H., Wang, J. and Steitz, T.A. (1996) Structure of Taq polymerase with DNA at the polymerase active site. *Nature* **382**(6588):278 – 81.
doi: http://dx.doi.org/10.1038/382278a0.

[24]   Altschul, S.F. *et al.* (1997) Gapped BLAST and PSI-BLAST: a new generation of protein database search programs. *Nucleic Acids Res.* **25**(17):3389 – 402.
doi: http://dx.doi.org/10.1093/nar/25.17.3389.

Experimental Standard Conditions of Enzyme Characterizations,
September 13th – 16th, 2009, Rüdesheim/Rhein, Germany

107

Different Contributions of the Various Isoenzymes to the Flux in the Aspartate-Derived Amino Acid Pathway in *Arabidopsis thaliana*

Gilles Curien[1], Renaud Dumas[1], Athel Cornish-Bowden[2] and María Luz Cárdenas[2,*]

[1]Laboratoire de Physiologie Cellulaire Végétale (LPCV), UMR 5168, CNRS – CEA – INRA – Université Joseph Fourier, Grenoble, France

[2]Unité de Bioénergétique et Ingénierie des Protéines (BIP), UPR 9036, CNRS, Marseille, France

E-Mail: *cardenas@ifr88.cnrs-mrs.fr

Received: 25th January 2010 / Published: 14th September 2010

Abstract

Since isoenzymes were first discovered, their physiological role has generated interest and discussion, and study of the flux distribution between isoenzymes in a real pathway, studied with real parameters, should shed light on this role. The aspartate-derived amino acid pathway from plants constitutes an excellent system for understanding the role of isoenzymes, as well as the effects of regulatory mechanisms such as feedback inhibition and allosteric interactions, because there are several branch-points, numerous isoenzymes, and different allosteric control mechanisms (inhibition, activation, antagonism and synergism). It is responsible for the distribution of the carbon flux from aspartate into the branches for synthesis of lysine, threonine, methionine and isoleucine. A mathematical model of the core of the pathway in the chloroplasts of *Arabidopsis thaliana* was constructed, and as kinetic data from the literature are often inadequate for kinetic modelling, we combined kinetic measurements obtained *in vitro* with purified enzymes, in near-physiological conditions, with *in vitro* reconstitution and numerical simulation. The model accurately predicts the experimentally observed behaviour, and shows that the isoenzymes contribute unequally to the flux and its regulation. The effects of some isoenzymes knockouts are also studied.

INTRODUCTION

The recognition by Markert and Møller half a century ago of multiple forms of enzymes catalysing the same reaction in the same cell or organism, i.e. isoenzymes, led to hopes of a better understanding of cellular metabolism through this phenomenon [1]. The discovery of the first isoenzymes thus raised the question of what physiological advantage derives from having more than one type of protein able to catalyse the same reaction. Different protein species can be generated by many different mechanisms, including different gene splicing. The term "isoenzyme" applies normally to proteins that are coded by different genetic loci, the evolutionary results of gene duplication or of gene duplication and fusion [2].

Isoenzymes tend to perform their functions in distinct ways, and they may differ in affinity and specificity for substrates or cofactors, in their response to allosteric effectors, subcellular localization, susceptibility to dietary and hormonal treatment, or time of appearance during differentiation [2]. The demonstration of one or several different properties can be used to explain the physiological advantages of a particular set of isoenzymes. For example, the different half-saturation values of the vertebrate hexokinases for glucose (and other kinetic parameters) may represent a regulatory device to handle overloads of dietary glucose [3, 4]. Especially intriguing are the differences in isoenzyme expression between related species. For example, although there are four hexokinase isoenzymes in rat liver (A, B, C, D), this is not a universal characteristic of mammalian liver or even of the livers of other rodents or of pig, which, like rat, is omnivorous. A system of three hexokinases is more common (ACB, ACD, ABD) and systems of two are also found (AB, AC) [5 – 7].

Studies on the metabolism of microorganisms revealed the existence of isoenzymes in the allosteric regulation of branched pathways. Their role in the synthesis of amino acids derived from aspartate was especially well investigated in *Escherichia coli*, particularly from the point of view of their allosteric properties [8]. At that time the global regulation of entire metabolic networks had not been clarified in a quantitative sense. It was assumed that there was a logical reason behind the existence in *E. coli* of three aspartokinases, each of them inhibited by a different amino acid product of the pathway. In some cases the same amino acid also repressed the corresponding gene expression.

One might be tempted by the idea that isoenzymes, especially if they have different properties, will be found at branch points of metabolic sequences. However, studies over many years show that isoenzymes are not restricted to highly regulated steps or to branch points in metabolism [9]. Reactions that are not at crossroads, and ones without obvious regulatory significance, are also catalysed by isoenzymes, and in several cases there are no differences in properties [2].

Recognition of kinetic differences within a group of isoenzymes should not be considered sufficient support on its own for speculations about the involvement of the isoenzymic system in regulation, even in the presence of a plausible correlation. However, the possibility of simulating a metabolic pathway offers a powerful tool for analysing the role of isoenzymes.

ASPARTATE METABOLISM IN CHLOROPLASTS OF *ARABIDOPSIS THALIANA*

The aspartate pathway of thale cress, *Arabidopsis thaliana*, provides an excellent opportunity for testing and developing ideas about metabolic regulation, the role of isoenzymes and the organization of metabolic pathways. In particular, it allows a test of one of the classical ideas of metabolic regulation, commonplace in textbooks, that the most highly regulated step is the one that controls the flux. There have been many theoretical discussions of such ideas [10 – 13] but almost none of them have been based on studies of real non-linear pathways, with quantitative experimental data obtained in physiologically relevant conditions.

The aspartate pathway is a branched pathway producing four amino acids (lysine, methionine, threonine and isoleucine) and in the reported study [14] we worked with a core of 13 enzymes. *Arabidopsis* has isoenzymes at several points, which respond unequally to effectors (Fig. 1). It has bifunctional enzymes and many different regulatory interactions, such as inhibition, activation, synergism and antagonism. Under physiological conditions the flow goes from aspartate to the other amino acids, but some reactions have equilibrium constants that favour the opposite direction. This is especially true for the reaction catalysed by aspartokinase (AK), where the equilibrium constant strongly favours the reverse reaction [15].

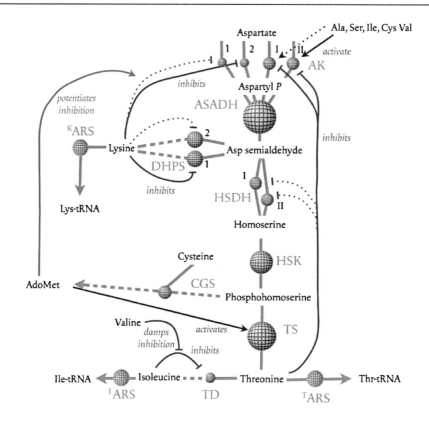

Figure 1. Model of aspartate metabolism in *Arabidopsis thaliana*. The pathway was modelled as a system of 13 enzymes (not including the three aminoacyl tRNA synthetases), as shown. The activation of AKI and AKII by five amino acids was not considered. Each turquoise sphere represents the concentration of the enzyme in question and should be considered three-dimensional, i.e. concentrations are proportional to the implicit volumes, not to the diameters, so ASADH, for example, has 46 times the activity of AK1. The activities of the three aminoacyl tRNA synthetases (yellow spheres) were treated as adjustable parameters.

Presence of isoenzymes

In the chloroplast there are four isoenzymes of aspartate kinase, which are designated in a most unfortunate way, as AK1, AK2, AKI and AKII. A fifth isoenzyme, AK3, exists but was not taken into account in the model because it is restricted to vascular tissues [16]. There are two isoenzymes of homoserine dehydrogenase, HSDH I and HSDH II; two of threonine synthase, TS 1 and TS 2, of which only TS 1 (referred to in the rest of the paper just as TS) exists in the chloroplast; and two of DHDPS, DHDPS 1 and DHDPS 2. As discussed below, all of these isoenzymes differ in substrate and effector affinity and the aspartate kinases also differ in effector specificity. A striking feature of this pathway is the presence of bifunctional enzymes: AKI and HSDH I are activities of the same protein, and the same applies to AKII and HSDH II. This bifunctionality has been conserved since the divergence of plants and bacteria, the two structural domains being quite independent [17].

Enzyme concentrations

The enzyme concentrations are not all the same: AKI and AKII are more abundant than AK1 and AK2, and there is 46 times as much ASADH as AK1, and about ten times as much as the four AK isoenzymes together.

Complexity of regulatory interactions

There are many and diverse regulatory interactions (Fig. 1). Lysine inhibits AK2 and DHDPS 1 strongly, and AK1 and DHDPS 2 weakly; however the weak inhibition of AK1 is highly potentiated by S-adenosylmethionine, the universal methyl donor, methionine as such exerting no regulatory interaction [18]. S-adenosylmethionine does not exert a direct feedback inhibition of its branch derived from phosphohomoserine, but it activates threonine synthase 1. The synergistic amplification by S-adenosylmethionine of the weak inhibition of AK1 by lysine allows demand for S-adenosylmethionine to regulate its production.

Isoleucine inhibits threonine deaminase, but valine damps this inhibition. Threonine in turn inhibits the two bifunctional aspartate kinases strongly, and the two homoserine dehydrogenases weakly; however HSDH I is more strongly inhibited than HSDH II, which is not at all inhibited by threonine in physiological conditions [17] The inhibition by threonine of AKI and AKII is counteracted by alanine, cysteine, isoleucine, serine and valine, which activate AKII strongly and AKI weakly [17]. Several of these effectors bind at ACT domains [19, 20]. Of all these effectors, alanine and threonine are much more abundant than the others in *Arabidopsis* leaf chloroplasts [17].

The complexity of this metabolism and of the regulatory interactions prevents an easy understanding of the function played by any given isoenzyme relative to another, and of the importance of the dual controls by lysine, threonine and S-adenosylmethionine.

There are several questions that can be asked. Which enzymes control the flux? Are these in the supply block (most of the enzymes) or the demand block (the aminoacyl tRNA synthetases)? How is the flux partitioned among the isoenzymes? Does every isoenzyme carry some flux? How well does the system regulate the flux in response to varying demand? How is the total flux distributed between the different branches? How independent are these branches? For the bifunctional enzymes, does AKI carry the same flux as HSDH I, and likewise for AKII and HSDH II? To answer these and other questions it is necessary to simulate the network as a whole.

MODELLING

For doing a suitable modelling a substantial amount of experimental data is required. Literature data are usually inadequate for kinetic modelling, because the kinetic parameters have not been obtained in the appropriate physiological conditions (pH, presence of products and effectors, etc.) [21], and often the precise conditions of measurement are not specified. Important parameters are often missing, especially those for the reverse reaction. If the equilibrium constant is very high and greatly favours the reaction in the physiological direction, the reverse reaction can be neglected, but inhibition by products must still be taken into account [22].

Fortunately, for the aspartate pathway in chloroplasts of *Arabidopsis* a considerable amount of information has accumulated over the years on the kinetic parameters of the different enzymes and on the physiological concentrations of enzymes and metabolites [17, 23 – 28]. This information allowed the construction of a detailed kinetic model for simulating and predicting the dynamic response of the pathway in conditions relevant to those *in vivo* [14].

To model a system such as the one represented in Figure 1, a system of equations is needed, together with a simulation program and, in particular, a great number of experimental measurements in conditions relevant to physiological conditions. As a core of 13 enzymes are considered, 13 catalytic constants, 13 enzyme concentrations, 63 other kinetic parameters, and 5 fixed concentrations (of "external" metabolites) are required, which means a total of 94 numerical values to be measured. This calculation suggests that it may be almost impossible to have a realistic simulation, but that is too pessimistic, as this type of modelling is typically very robust, in the sense that two-fold or even larger errors in the values of the parameters typically have little quantitative effect on the predictions, and no qualitative effect.

The existence of two types of aspartate kinases, two isoenzymes specifically evolved for synthesis of lysine and maybe methionine (AK1, AK2) and two (AKI, AKII) for the synthesis of threonine and isoleucine may suggest the possibility of channelled routes to these amino acids. However, there is no evidence of channelling, and all indications are that there is a single pool of aspartyl-phosphate derived from the action of the four aspartate kinase isoenzymes, and this is what was assumed in the modelling.

Concentrations of external metabolites

The values of [aspartate] = 1.5 mM, [cysteine] = 15 μM, [s-adenosylmethionine] = 20 μM and [valine] = 100 μM are those measured experimentally and are treated as constant during the time where the model is valid (2 hours). The same applies to the concentrations of ATP, ADP, NADPH, NADP, inorganic phosphate and pyruvate, which are considered fixed concentrations because their levels are assumed to be regulated elsewhere in the metabolism.

$$v_{AK1} = [AK1] \cdot \frac{5.65 - 1.6[AspP]}{1 + \left[[Lys]/\left(\dfrac{550}{1 + [AdoMet]/3.5}\right)\right]^2}$$

$$v_{HSK} = [HSK] \cdot \frac{\left(\dfrac{2.85[ATP]}{54 + [ATP]}\right) \cdot [Hser]}{12 + \dfrac{40}{1 + [ATP]/80} + [Hser]}$$

$$v_{TS1} = [TS1] \cdot \frac{\left(\dfrac{0.42 + 3.5[AdoMet]^2/73}{1 + [AdoMet]^2/73}\right)[PHser]}{\left[250 \dfrac{\left(\dfrac{1 + [AdoMet]/0.5}{1 + [AdoMet/1.1}\right)}{1 + \dfrac{[AdoMet]^2}{140}}\right]\left(1 + \dfrac{[P_i]}{1000}\right) + [PHser]}$$

Figure 2. Illustration of the rate equations required for the modelling. The rate expressions for AK1, HSK and TS1 are shown. Corresponding expressions were also required for the other steps in the system.

Kinetic equations

The three equations shown in Figure 2, which correspond to AKI, HSK and TS 1, illustrate the sort of equations required for modelling the system. They are not mechanistic equations, but empirical ones derived from experimental data, and validated in reconstituted experiments *in vitro*. Each of the first two equations requires 5 measured parameters (including the Hill coefficient for AK1, $h = 2$,) but TS 1 requires no fewer than 12 parameters.

Internal metabolites

The concentrations of metabolites aspartyl-*p*, aspartate semialdehyde, homoserine, lysine, phosphohomoserine, threonine and isoleucine were calculated on the basis of the kinetic equations of the enzymes by using the simulation program COPASI [28]. All the fluxes were calculated at the same time as the internal concentrations. The calculated internal metabolite concentrations are, within experimental error, the same as those measured in chloroplast stroma in light conditions, for those concentrations that were high enough to allow detection. The low calculated concentration of aspartyl-*p* agrees with the fact that this concentration is below the level of experimental detection.

FLUX DISTRIBUTION IN THE REFERENCE STATE

The model allowed us to calculate the flux through the different branches of the network and to establish how much flux is carried by each isoenzyme. The analysis refers to a state that we define as the *reference state*, which corresponds to the metabolite and enzyme concentrations measured in chloroplast stroma in light conditions (pH 8.0). It showed a different flux distribution between the different isoenzymes and in branch points. The flux tends to be unequally distributed between the different isoenzymes and in various respects is quite different from what one might expect: For this sort of system intuition is not sufficient and can be misleading. At the step of aspartate phosphorylation the lysine-sensitive AK1 and AK2 account for most of the flux (88%), while the threonine-sensitive AKI and AKII account for very little, AK1 being the one that carries most of the flux (73%). There is also an important difference between the dihydrodipicolinate synthase isoenzymes: DHDPS 2 accounts for 28% of the total flux whereas DHDPS 1, which is strongly inhibited by lysine, carries only 3%.

Contrary to what intuition would suggest, the two activities of each of the bifunctional enzymes AKI-HSDH I and AKII-HSDH II, carry very different amounts of flux: 32% for HSDH I but only 3% for AKI, 37% for HSDH II but only 9% for AKII. Notice that there is no major difference between the fluxes carried by the two homoserine dehydrogenase isoenzymes.

Another striking feature is the fact that at the bifurcation point at phosphohomoserine the flux to *S*-adenosylmethionine, 0.058 µM/s (5.7% of the total flux) is much smaller than the flux to threonine (63%). The activation of TS by *S*-adenosylmethionine probably avoids a "branch-point effect" [29] whereby small changes in the synthesis of threonine would perturb the regulation of *S*-adenosylmethionine production, but the simulations themselves did not deal with that because *S*-adenosylmethionine was defined as an external metabolite at fixed concentration. The model explains why despite the activation of TS by *S*-adenosylmethionine an inverse relationship between its concentration and that of threonine was observed experimentally.

The demand for amino acids (in the form of the activities of the three AA tRNA synthetases) was adjusted to obtain concentrations of intermediate metabolites close to those measured *in vivo* (Reference state). Metabolic control analysis [10, 30, 31] was used to identify the most sensitive steps and metabolites of the network. Most of the control of the common flux is shared between the 3 AA-tRNA synthetases, which represent the cellular demand for the amino acids, and not at the highly regulated AK step, confirming the hypothesis of Kacser and Burns [10], later developed in detail [11], that the effect of feedback inhibition is to *withdraw* the control from the first step and to transfer it to the demand. This is important as an illustration that controls and regulation are not the same thing. However, in the reference state there is still some control associated with AK1, which is at the "supply" end of the pathway, with the consequence that AK1 contributes significantly to the maintenance of the steady state of threonine. This is highlighted in the knockout simulations analysed below.

HOW WELL IS THE SYSTEM REGULATED IN PRACTICE?

If lysine is consumed in substantial amounts by competing pathways does that make it less available for making protein? The modelling showed that the flux of threonine to protein is not affected at all in this condition and the flux from lysine to protein is affected only slightly. There is an equivalent result if threonine is consumed in substantial amounts by competing pathways: the flux of lysine to protein is not affected at all and the flux of threonine to protein is affected only slightly. We can ask the same question in relation to methionine by modulating the concentration of S-adenosylmethionine, with a similar answer. The system thus appears to be very robust in relation to *increases* in demand. However, if the demand for the three end-products is too low the system is not able to cope and there is no steady state. This result is interesting in the light of the results obtained with knockouts of the three AA-tRNA synthetases, shown below. Although amino acid catabolism was not explicitly included in the model the simulations of high demand for threonine and lysine suggest that the results would not probably be affected qualitatively if it had been.

It follows that the different branches of the network respond independently to fluctuations in the demand for their end-products, even though they share common steps. The kinetic independence of the two domains of the bifunctional AK-HSDHs, and the different response of isoenzymes to effectors contributes to this independence. However, as discussed below, knockouts in one branch can affect enormously the metabolite concentrations of others.

KNOCKOUTS IN SILICO

With the idea of shedding some light in relation to the physiological role of isoenzymes, in particular of those that carry almost no flux, some knockout experiments were done *in silico*, as illustrated in Table 1. It is striking that a knockout of AK1 still allows 80% of the total flux, even though this isoenzyme carries 73% of the flux in the reference state. In this condition, therefore, the other three isoenzymes can compensate for the knockout. Even if both lysine-sensitive aspartate kinases are eliminated, leaving only AKI and AKII active, two isoenzymes that account only 12% of the flux in the wild type, the resulting total flux is still 74% of that in the wild type: The two threonine-sensitive isoenzymes can compensate for loss of the other two.

Table 1. Effects of knock-outs on fluxes and intermediate concentrations. The top line ("wild-type") shows the common flux (through aspartate semialdehyde dehydrogenase) and the concentrations of threonine, lysine and isoleucine in the reference state, and the one rows show the same variables when the activities of the enzyme or enzymes in the left-hand column are set to zero

Step	JAK = JASADH	[Thr]	[Lys]	[He]
Wild type	1.016	296	69.2	58.8
AK1	0.812	92.7	54.6	32.9
AK2	0.974	223	65.7	52.3
AK1 and 2	0.749	70.3	50.6	26.4
AK2, I and II	0.869	122	58.2	39.2
AKI	1.010	285	68.8	58.0
AKII	0.978	228	66.0	52.9
AK I and II	0.869	206	64.8	50.6
DHDPS 1	1.025	330	66.8	61.3
DHDPS 2	1.141	7760	37.9	179.2
DHSPS 1 and 2	- no steady state -			
Lys^{tRNA}	0.543	104	5×10^8	35.5
Thr^{tRNA}	0.757	11400	81.9	203
Ile^{tRNA}	0.789	6600	79.0	2×10^{13}

It is important to notice that although the flux through aspartic semialdehyde dehydrogenase is not very much affected if AK1 is knocked out, the threonine concentration is greatly affected, as it falls to only 31% of the wild-type value with the AK1 knockout, and to 24% with the double knockout: this illustrates the general property that metabolite concentrations are more susceptible to perturbations than fluxes [22, 32]. This decrease in the threonine concentration relieves the inhibition of AKI and AKII, allowing them to replace AK1 and AK2.

For the dihydrodipicolinate synthase isoenzymes, the result obtained with the knockout of DHDPS2, the isoenzyme weakly inhibited by lysine, is striking: there is a slight increase in the total flux (to 112%), which may be explained by the decrease in the concentration of lysine, which is now 55% of the reference value. Conversely, the isoleucine and threonine concentrations are dramatically increased, isoleucine to 305% and threonine to 2600%, in agreement with the observation that the concentration of threonine is very unstable [14]. So although the amount of lysine would probably still be sufficient to satisfy the plant needs, the huge increase in threonine concentration would be toxic. This result illustrates the point that one needs to be cautious in relation to the interpretation of knockout effects, as pointed out by Cornish-Bowden and Cárdenas [33]: In this case the damaging effect is not so much the decrease in lysine but the huge increase in threonine. As could be expected, knockout of both isoenzymes of dihydrodipicolinate synthase prevents a steady state from being attained. However, as note below, the model does not include catabolism, and in the living plant this loss of steady state might not occur.

CONCLUSIONS AND PERSPECTIVES

There have been some excellent simulations of metabolic systems, such as erythrocyte metabolism [34], or glycolysis in *Trypanosoma brucei* [35], and sucrose metabolism in sugarcane [36] but these have not addressed the role of feedback inhibition in branched pathways. Studies of feedback inhibition have, on the other hand, referred to models of invented pathways with hypothetical data [11, 12].

For the first time, therefore, we can describe metabolic regulation in a detailed model of a real system with measured kinetic parameters. Most of the control of the network is on the demand steps (AA-tRNA synthetases), and not at the severely regulated AK, confirming that the effect of feedback inhibition is to transfer the control from the first step of a pathway to the demand steps [10,11], and showing the difference between control and regulation.

The presence of isoenzymes makes the system very robust and contributes to the constancy of the flux; enzymes that carry almost no flux in the reference state can take care of it if another isoenzyme is missing. Although the branches appear independent in relation to variations in demand for the different amino acids, because the fluxes are not disturbed, the steady state concentrations of certain metabolites, such as threonine, could be highly disturbed in the absence of certain isoenzymes. In addition, the existence of isoenzymes could be seen as an evolutionary strategy for being able to respond to a wide range of effectors, which would be more difficult with just one enzyme, in which case a partial effect of effectors is required (cumulative inhibition or activation). The alternative for allowing the action of several different effectors being regulation by covalent modification (metabolic cascades), which in addition allows a higher sensitive response.

The system appears to have evolved to cope with increases in demand, as the demand for lysine or threonine can increase considerably without major perturbation [14]. However, decreases in demand below a certain threshold considerably altered metabolite concentrations, as illustrated in Table 1 with the knockouts at the demand step, and could even prevent a steady state from being attained. Notice, however, that the simulated knockout studies do not consider amino acid catabolism, and in the living plant it is possible that catabolic reactions would prevent the huge changes of concentrations (especially of threonine) that we see in the simulations. It will be interesting to incorporate this in future studies, as well as the interactions that may exist with other amino acid pathways, such as amide amino acid metabolism and pyruvate metabolism. The latter is important, because part of the isoleucine molecule derives from pyruvate, and as this is also the case for valine it may be related to the regulatory effect of valine on the inhibition of threonine deaminase by isoleucine, which appears puzzling if considered in isolation.

As pointed out by Mazat and Nazaret [37] this modelling of the global network of the branched pathway derived from aspartate [14] should not be seen as an end. On the contrary, it should provide a basis for the interpretation of changes that can be produced in the plant. Furthermore it will provide a framework to extend similar simulations to other organisms; it would be interesting if the study done in *E. coli* in relation to the control of threonine synthesis [38] could be extended.

ABBREVIATIONS

AK, aspartate kinase;
HSDH, homoserine dehydrogenase;
DHDPS, dihydrodipicolinate synthase;
ASADH, aspartic semialdehyde dehydrogenase;
TD, threonine deaminase;
TS, threonine synthase;
AA-tRNA synthetase, aminoacyl-tRNA synthetase

REFERENCES

[1]   Markert, C.L., Møller, F. (1959) Multiple forms of enzymes: Tissue, ontogenetic, and species specific patterns. *Proc. Natl. Acad. Sci. U.S.A.* **45**:753–763.
       doi: http://dx.doi.org/10.1073/pnas.45.5.753.

[2]   Ureta, T. (1978). The role of isozymes in metabolism: a model of metabolic pathways as the basis for the biological role of isozymes. *Curr. Top. Cell. Reg.* **13**:233–258.

[3]   Niemeyer, H., Ureta, T., Clark-Turri L. (1975) Adaptive character of liver gluco-kinase. *Mol. Cell. Biochem.* **6**:109–126.
       doi: http://dx.doi.org/10.1007/BF01732005.

[4]   Cárdenas, M.L., Cornish-Bowden, A., Ureta, T. (1998) Evolution and regulatory role of the hexokinases. *Biochim. Biophys. Acta* **1401**:242–264.
       doi: http://dx.doi.org/10.1016/S0167-4889(97)00150-X.

[5]   Ureta, T., Radojkovic, J., Zepeda, S. Guixé, V. (1981) Comparative studies on glucose phosphorylating isoenzymes of vertebrates. VII Mammalian hexokinases. *Comp. Biochem. Physiol.* **70B**:225–236.

[6]   Ureta, T. (1982) The comparative isozymology of vertebrate hexokinases. *Comp. Biochem. Physiol.* **71B**:549–555.

[7]   Cárdenas, M.L. (2004) Comparative biochemistry of glucokinase. In: *Glucokinase and Glycemic Disease: from Basics to Novel Therapeutics* (ed. F. M. Matschinsky & M. A. Magnuson), Karger, Basle, pp. 31–41.

[8]   Patte, J.C., Loviny, T., Cohen, G.N. (1965) Co-operative inhibitory effects of L-lysine with other amino acids on an aspartokinase from *Escherichia coli. Biochim. Biophys. Acta* **99**:523 – 530

[9]   Markert, C.L. ed. (1975), *Isozymes*, 4 vols. Academic Press, New York.

[10]   Kacser, H., Burns, J.A. (1973) The control of flux. *Symp. Soc. Exp. Biol.* **32:** 65 – 104.

[11]   Hofmeyr, J.-H. S., Cornish-Bowden, A. (1991) Quantitative assessment of regulation in metabolic systems. *Eur. J. Biochem.* **200**:223 – 236.
       doi: http://dx.doi.org/10.1111/j.1432-1033.1991.tb21071.x.

[12]   Cornish-Bowden, A., Hofmeyr, J.-H. S., Cárdenas, M. L. (1995) Strategies for manipulating metabolic fluxes in biotechnology. *Bioorg. Chem.* **23**: 439 – 449.
       doi: http://dx.doi.org/10.1006/bioo.1995.1030.

[13]   Fell, D. (1997) *Understanding the control of metabolism*. Portland Press, London.

[14]   Curien, G., Bastien, O., Robert-Genthon, M., Cornish-Bowden, A., Cárdenas, M.L., Dumas, R. (2009) Understanding regulation of aspartate metabolism with a model based on measured kinetic parameters. *Mol. Syst. Biol.* **5**:271.
       doi: http://dx.doi.org/10.1038/msb.2009.29.

[15]   Shaw, J.F., Smith, W.W. (1977) Studies on the kinetic mechanism of lysine-sensitive aspartokinase. *J. Biol. Chem.* **252**:5304 – 5309.

[16]   Yoshioka, Y., Kurei, S., Machida, Y. (2001) Identification of a monofunctional aspartate kinase gene of *Arabidopsis thaliana* with spatially and temporally regulated expression. *Genes Genet. Syst.* **76**:189 – 198.
       doi: http://dx.doi.org/10.1266/ggs.76.189.

[17]   Curien G., Ravanel S. Robert M., Dumas R. (2005) Identification of six novel allosteric effectors of *Arabidopsis thaliana* aspartate kinase-homoserine dehydrogenase isoforms. Physiological context sets the specificity. *J. Biol. Chem.* **280**:41178 – 41183.
       doi: http://dx.doi.org/10.1074/jbc.M509324200.

[18]   Curien G., Laurencin M., Robert-Genthon M., Dumas R. (2007) Allosteric monofunctional aspartate kinases from *Arabidopsis. FEBS J.* **274**:164 – 176.
       doi: http://dx.doi.org/10.1111/j.1742-4658.2006.05573.x.

[19]   Mas-Droux C., Curien G., Robert-Genthon M., Laurencin M., Ferrer, J.-L., Dumas R. (2006) A novel organization of ACT domains in allosteric enzymes revealed by the crystal structure of *Arabidopsis* aspartate kinase. *Plant Cell* **18**:1681 – 1692.
       doi: http://dx.doi.org/10.1105/tpc.105.040451.

[20]   Curien, G, Biou, V., Mas-Droux, C., Robert-Genthon, M., Ferrer, J.L., Dumas, R. (2008) Amino acid biosynthesis: new architectures in allosteric enzymes. *Plant Physiol Biochem.* **46**:325 – 339.
doi: http://dx.doi.org/10.1016/j.plaphy.2007.12.006.

[21]   Cornish-Bowden, A., Hofmeyr, J.-H. S. (2005) Enzymes in context: kinetic characterization of enzymes for systems biology. *The Biochemist* **27**:11 – 14.

[22]   Cornish-Bowden, A., Cárdenas M.L. (2001) Information transfer in metabolic pathways: effects of irreversible steps in computer models. *Eur. J. Biochem.* **268**:6616 – 6624.
doi: http://dx.doi.org/10.1046/j.0014-2956.2001.02616.x.

[23]   Krause, G.H., Heber, U. (1976) Energetics of intact chloroplasts. In: *The intact chloroplast* (ed. Barber, J.), pp 174 – 175. Elsevier/North-Holland Biomedical Press, The Netherlands

[24]   Bligny, R., Gardestrom, P., Roby, C., Douce, R. (1990) ^{31}P NMR studies of spinach leaves and their chloroplasts. *J. Biol. Chem.* **265**:1319 – 1326

[25]   Curien, G., Job, D., Douce, R., Dumas R. (1998) Allosteric activation of *Arabidopsis* threonine synthase by *S*-adenosylmethionine. *Biochemistry* **37**:13212 – 13221.
doi: http://dx.doi.org/10.1021/bi980068f.

[26]   Curien G., Ravanel, S., Dumas, R. (2003) A kinetic model of the branch-point between the methionine and threonine biosynthesis pathways in *Arabidopsis thaliana*. *Eur. J. Biochem.* **270**:4615 – 4627.
doi: http://dx.doi.org/10.1046/j.1432-1033.2003.03851.x.

[27]   Paris, S., Viemon, C., Curien, G., Dumas, R. (2003) Mechanism of control of *Arabidopsis thaliana* aspartate kinase–homoserine dehydrogenase by threonine. *J. Biol. Chem.* **278**:5361 – 5366.
doi: http://dx.doi.org/10.1074/jbc.M207379200.

[28]   Hoops, S., Sahle, S., Gauges, R., Lee, C., Pahle, J., Simus, N., Singhal, M., Xu, L., Mendes, P., Kummer, U. (2006) COPASI-a COmplex PAthway SImulator *Bioinformatics* **22**:3067 – 3074.

[29]   LaPorte, D.C., Walsh, K., Koshland, D., Jr. (1984) The branch point effect: ultrasensitivity and subsensitivity to metabolic control. *J. Biol. Chem.* **259**:14068 – 14075.

[30]   Heinrich, R., Rapoport, T.A. (1974) A linear steady-state treatment of enzymatic chains. General properties, control and effector strength. *Eur. J. Biochem.* **42**:89 – 95.
doi: http://dx.doi.org/10.1111/j.1432-1033.1974.tb03318.x.

[31] Reder, C. (1988) Metabolic control theory: a structural approach. *J. Theor. Biol.*
      135:175 – 201.
      doi: http://dx.doi.org/10.1016/S0022-5193(88)80073-0.

[32] Cornish-Bowden, A., Cárdenas, M.L. (2001) Silent genes given voice. *Nature*
      409:571 – 572.
      doi: http://dx.doi.org/10.1038/35054646.

[33] Cornish-Bowden, A., Cárdenas, M.L. (2001) Complex networks of interactions con-
      nect genes to phenotypes. *Trends Biochem. Sci.* **26**:463 – 465.
      doi: http://dx.doi.org/10.1016/S0968-0004(01)01920-X.

[34] Mulquiney, P.J., Kuchel, P.W. (1999) Model of 2,3-bisphosphoglycerate metabolism
      in the human erythrocyte based on detailed enzyme kinetic equations: computer
      simulation and metabolic control analysis. *Biochem. J.* **342**:597 – 604.
      doi: http://dx.doi.org/10.1042/0264-6021:3420597.

[35] Helfert, S., Estévez, A.M., Bakker, B., Michels, P., Clayton, C. (2001) Roles of
      triosephosphate isomerase and aerobic metabolism in *Trypanosoma brucei*.
      Biochem. J. **357**:117 – 125.
      doi: http://dx.doi.org/10.1042/0264-6021:3570117.

[36] Uys, L., Botha, F.C., Hofmeyr, J.H., Rohwer, J.M. (2007) Kinetic model of sucrose
      accumulation in maturing sugarcane culm tissue. *Phytochemistry* **68**:2375 – 2392.
      doi: http://dx.doi.org/10.1016/j.phytochem.2007.04.023.

[37] Mazat J.P., Nazaret C. (2009) Branches in the plant world. *J. Biosci.* **34**:343 – 344.
      doi: http://dx.doi.org/10.1007/s12038-009-0038-y.

[38] Chassagnole, C., Fell, D.A., Raÿs, B., Mazat J.-P. (2001) Control of the threonine
      synthesis pathway in *Escherichia coli*: a theoretical and expérimental approach.
      Biochem. J. **35**:4333 – 4344.

Parameterization of Large-Scale Autonomous Network Models from Time-Series Metabolite Data

Klaus Mauch[1,*], Ute Hofmann[2], Matthias Reuss[3], Klaus Maier[1,3]

[1]Insilico Biotechnology AG, Nobelstrasse 15, 70569 Stuttgart, Germany

[2]Dr. Margarete Fischer-Bosch Institute of Clinical Pharmacology, Stuttgart and University of Tuebingen, Auerbachstrasse 112, 70376 Stuttgart, Germany

[3]Institute of Biochemical Engineering, University of Stuttgart, Allmandring 31, 70569 Stuttgart, Germany

E-Mail: *klaus.mauch@insilico-biotechnology.com

Received: 17th June 2010 / Published: 14th September 2010

Abstract

This contribution illustrates a framework for reconstructing and verifying large-scale kinetic metabolic network models in a semi-automated way. The experimental basis of this approach is provided by a stimulus response experiment. Parameterizing a large-scale kinetic network model then consists of three steps. In a first step, metabolic fluxes are identified. Subsequently, the dynamic network model is set-up automatically using canonical enzyme kinetics. In a third step, the kinetic parameters of the model are estimated from time-series metabolite data through integrating an evolutionary algorithm with a high-performance dynamic simulation platform. In this study, the time-series metabolite data were collected from HepG2 cells and analysed in the range of 0 to 180 minutes after depriving glucose from the culture medium. In total, more than 6 million simulation runs were performed. The *in silico* metabolite dynamics were in accordance with the experimental data.

INTRODUCTION

Kinetic network models allow for quantitative and systems-oriented predictions of cellular productivities, detoxification processes, and metabolic disorders like, for example, diabetes. Moreover, dynamic network models might eventually enable the personalized prognosis of drug actions and/or their persistency. However, to achieve a broad application and acceptance of dynamic *in silico* cells in the life science industries, a number of challenges need to be overcome. First and foremost, the model building currently is (too) slow and (too) costly. Next, the model representation and validation lack of standardized quality criteria, while setting up and managing large-scale network models lack of efficient computational tools. Finally, despite remarkable recent advances in metabolomics, acquiring adequate experimental data is still challenging, i. e. the issues of (fast) sampling and (reliable) quenching of single cells, cellular compartments and/or tissue are far from resolved.

In order to overcome the outlined challenges and to accelerate the reconstruction and identification of large-scale dynamic network models, the workflow shown in Figure 1 is proposed in this contribution.

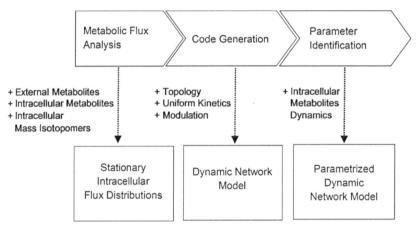

Figure 1. Workflow for parameterization of enzyme kinetics.

The complexity of the model parameterization process is effectively mitigated by breaking it down into a three-step procedure. In a first step, metabolic fluxes are identified. Subsequently, the dynamic network model is set-up automatically using canonical enzyme kinetics. In a third step, the kinetic parameters of the model are then estimated from time-series metabolite data. Identifying the network dynamics by referring to a physiologically relevant flux distribution confines both parameter range and network dynamics significantly. Moreover, the predictive quality of autonomous network models improves the closer predicted states to the experimentally observed reference state.

In the present study, a stimulus response experiment was performed using hepatoma cells. After growing HepG2 cells on a glucose-containing medium, the cells were incubated in fresh medium for two hours and then exposed to a medium lacking glucose. Metabolite time-series data were taken and analysed in the range of 0 to 180 minutes. The following sections will focus on (i) experimental methods, (ii) reconstruction of a large-scale autonomous network model, (iii) parameterization of the network model, and (iv) verification of *in silico* results.

EXPERIMENTAL METHODS

HepG2 cells (ATCC® Number HB-8065™) were incubated at 37 °C in 6-well-plates in 5% CO_2 atmosphere. The cells were cultured in alanyl-glutamine-free William's medium E (PAN Biotech GmbH, Aidenbach, Germany) which was supplemented with penicillin (100 U/mL), streptomycin (100 mg/mL), and GibcoTM Insulin-Transferrin-Selenium (100X) supplement (Invitrogen, Karlsruhe, Germany). No fetal calf serum was added to the medium. The 6-well plates were shaken at 20 rpm throughout the experiment (Shaker DRS-12, ELMI, Riga, Latvia). The number of cells was determined using a Neubauer counting chamber. The intracellular flux map corresponding to this experimental setup was determined previously [1, 2]. The main flux was found to be the conversion of glucose to lactate. Thus, for designing an efficient stimulus response experiment, the glucose flux was considered as the most promising candidate for perturbing the central metabolism of the hepatoma cells. However, the cells were grown in a batch culture, and extracellular glucose was provided in excess. Therefore, it was concluded that an extracellular glucose pulse would not yield essential changes, whereas glucose deprivation was expected to trigger a substantial metabolic response.

Before depriving the cells of extracellular glucose, they were treated as described in [1]: The overnight medium was replaced with fresh culture medium, which was then exchanged with glucose-free medium after 2 h of equilibration. Extra- and intracellular samples were collected in triplicate directly before and after the stimulus, as well as 1, 2, 5, 10, 30, 60, 120, and 180 min after glucose deprivation. The sampling approach and the processing of the samples were done as described by Hofmann et al. [1].

The concentrations of alanine, serine, glucose, lactate, pyruvate, fumarate, malate, cis-aconitate, isocitrate, and citrate were determined by GC-MS as described in [1, 2]. After glucose deprivation, the extracellular glucose concentrations were determined in 10 µl of diluted (1:9 v/v) medium samples, the intracellular glucose concentrations before and after perturbation were determined in 5 and 50 µl of cell extract, respectively. Phosphoenolpyruvate, 3-phosphoglycerate, dihydroxyacetonphosphate, fructose-1,6-bisphosphate, glucose-6-phosphate, 6-phosphogluconate, sedoheptulose-7-phosphate, ribose-5-phosphate, and ribulose-5-phosphate were determined by LC-MS-MS as described by Hofmann et al. [1] with the following modifications: HPLC separation was performed at 20 °C on a Synergi Hydro-RP column

(150 x 2 mm, 4 µm; Phenomenex, Aschaffenburg, Germany) at a flow rate of 0.2 ml/min. The mobile phases consisted of (A) water with 10 mM tributylamine and 15 mM acetic acid, and (B) methanol. Gradient runs were programmed as follows: 100% A from 0 to 10 min, increase to 20% B to 25 min, remaining at 20% B to 30 min, increase to 35% B to 35 min, remaining at 35% B to 40 min, increase to 60% B to 45 min, increase to 90% B to 48 min remaining at 90% B to 50 min, then equilibrating with 100% A for 13 min. Precursor and product ions used for the quantification of glucose-6-phosphate, 6-phosphogluconate, ribose-5-phosphate, ribulose-5-phosphate, fructose-1,6-bisphosphate, and the internal standard mannitol-1-phosphate were as previously described [1] and for phosphoenolpyruvate: m/z 167/97, 139; 3- phosphoglycerate: m/z 185/97, 167; dihydroxyacetonphosphate: m/z 169/ 97 and sedoheptulose-7-phosphate: m/z 289/97, 199.

Nucleotide analysis was performed by reversed phase ion pair high performance liquid chromatography. The HPLC system (Agilent Technologies, Waldbronn, Germany) consisted of an Agilent 1200 series autosampler, an Agilent 1200 series Binary Pump SL, an Agilent 1200 series thermostated column compartment, and an Agilent 1200 series diode array detector set at 260 and 340 nm. The nucleotides were separated and quantified on an RP-C-18 column that was combined with a guard column (Supelcosil LC-18-T; 15 cm x 4.6 mm, 3 µm packing and Supelguard LC-18-T replacement cartridges, 2 cm; Supelco, Bellefonte, U.S.A.) at a flow rate of 1 ml/min. The mobile phases were (i) buffer A (0.1 M KH_2PO_4/ K_2HPO_4, with 4 mM tetrabutylammonium sulfate and 0.5% methanol, ph 6.0) and (ii) solvent B (70% buffer A and 30% methanol, pH 7.2). The following gradient programs were implemented: 100% buffer A from 0 min to 3.5 min, increase to 100% B until 43.5 min, remaining at 100% B until 51 min, decrease to 100% A until 56 min and remaining at 100% A until 66 min.

Network Reconstruction

A kinetic network model was set up that includes the core metabolic pathways of hepatoma cells. The network model is based on a previously published isotopomer model that had been used for the estimation of intracellular fluxes from transient ^{13}C-labeling data [3] and accounts for 45 balanced compounds that are converted into each other by 49 reactions, including 5 transportation steps. The corresponding metabolic scheme is shown in Figure 2.

Parameterization of large-scale autonomous network models from time-series metabolite data

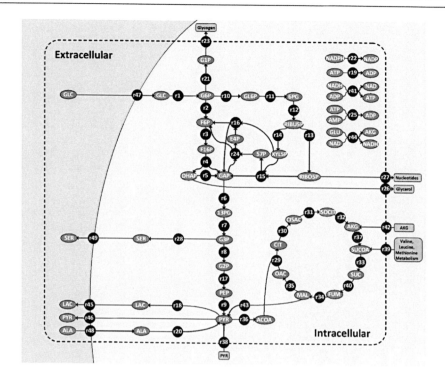

Figure 2. Metabolic network model of the hepatoma central metabolism. Extra- and intracellular metabolites are depicted with blue ellipses. Enzymatic reactions and transportation steps are indicated with red circles. Non-balanced compounds are shown within grey rounded rectangles. Directions of arrows reflect the direction of the steady state fluxes. The system boundary is indicated with a dashed line.

In the context of oxidative phosphorylation and the dynamic interplay of catabolism and anabolism, the cofactors NAD(H), NADP(H), ATP/ADP/AMP need to be taken into account by mass balances when analyzing the systems-level effect of the energy metabolism. The metabolic pathways under consideration contain 3 conserved moieties ($c_{amp}+c_{adp}+c_{atp}=$ const; $c_{nadp}+c_{nadph}=$ const; $c_{nad}+c_{nadh}=$ const). The model comprises glycolysis (EMP), the pentose-phosphate pathway (PPP), and the tricarboxylic acid (TCA) cycle. In the cataplerotic section, the malic enzyme, which decarboxylates malate to pyruvate, is taken into account. Reduced NADH is regenerated in the lactate dehydrogenase and oxidative phosphorylation reactions. P/O ratios of 2.5 and 1.5 were assumed for NADH and succinate, respectively [4]. Based on experimental evidence [3], the metabolic state was assumed to be that of fed hepatic cells. Accordingly, no gluconeogenetic reactions were included. Exchange fluxes with the system boundary took into account glucose and alanine uptake, glycogen storage, the metabolism of glutamate, valine, leucine, and methionine, glycerol and nucleotide synthesis, as well as serine, lactate, and pyruvate excretion. In addition, reactions that represented ATP and NADPH consumption relating to the basal metabolism were included. 31 regulatory effects (21 inhibitions and 10 activations) were found in a literature search [5] and included in the model (cf. Table 1).

Table 1. Activator and inhibitor influences. Regulatory influences and corresponding literature references. The modulator effects were included in the dynamic network model. In addition to these regulatory influences, the dynamic model did also account for substrate and product effects.

Enzyme Identifier	EC-Number	Activators	Inhibitors
glucokinase	2.7.1.2		F6P [23]
glucose-6-phosphate isomerase	5.3.1.9		6PG [24]
6-phosphofructokinase	2.7.1.11	AMP [25]	CIT [26]
fructose-bisphosphate aldolase	4.1.2.13		ADP, ATP, E4P, F6P, G1P, G6P, RIBO5P [27]
triose-phosphate isomerase	5.3.1.1		ATP [28]
glyceraldehyde-3-phosphate dehydro-genase	1.2.1.12		ADP, ATP [29]
phosphoglycerate kinase	2.7.2.3		AMP [30]
pyruvate kinase	2.7.1.40	G6P, F6P, G1P [31], F16P [32]	ALA [33]
glucose-6-phosphate dehydrogenase	1.1.1.49		ATP [34]
phosphoglucomutase	5.4.2.2		F16P [35]
UTP-glucose-1-phosphate uridylyltrans-ferase	2.7.7.9		AMP [36]
alpha-ketoglutarate dehydrogenase	1.2.4.2	ADP [37]	ATP [37]
valine, isoleucine, methionine metabo-lism	-	NAD, AKG [38]	GLU, NADH [39]
isocitrate dehydrogenase	1.1.1.41	ADP [40]	
pyruvate dehydrogenase	1.2.4.1	AMP [41]	

The network model discriminated between 5 extracellular (glucose, lactate, serine, pyruvate, alanine) and 40 intracellular metabolites. The metabolic pathways neither contained dead-end metabolites nor strictly detailed balanced sub-networks [6]. Furthermore, all reactions were consistent with respect to mass conservation and redox state.

The following set of metabolite mass balances was set-up automatically to describe the time-dependent behaviour of a metabolic system:

$$\frac{d}{dt}\left(\frac{c}{c^0}\right) = \left(c^0\right)^{-1} \cdot N \cdot r \tag{1}$$

N denotes the stoichiometry matrix and r the rate vector. (c^0) is a square diagonal matrix with reference concentrations on its main diagonal; $\left(\frac{c}{c^0}\right)$ denotes the normalized metabolite concentration vector.

Canonical linear-logarithmic (linlog) kinetics was automatically assigned for approximating the reaction rates in equation (1) [7 – 9]. The linlog formalism has been used for modelling *in vivo* kinetics and metabolic redesign [7]. Linlog kinetics was shown to have a good approximation quality and to need only relatively few parameters to be identified [10 – 12]. In linlog kinetics, all rate equations share a standardized mathematical format in which influences of metabolite and effector levels on reaction rates are taken into consideration by adding up logarithmic concentration terms. The standardized format is highly advantageous with a view to automating and speeding up the model set-up process. Besides, even for well-studied pathways of the central carbon metabolism it is often not the case that all kinetic mechanisms are known in detail [13, 14]. The matrix notation of the linlog rate equation is given by [7]

$$r = J^0 \cdot \left(\frac{e}{e^0}\right) \cdot \left(i + E_c^0 \cdot ln\left(\frac{c}{c^0}\right)\right) \tag{2}$$

in which J^0 is the reference steady state flux distribution, $\left(\frac{e}{e^0}\right)$ is a diagonal matrix containing relative enzyme levels, and i is a vector of ones. E_c^0 is a matrix whose entries are scaled elasticity coefficients ε_{ij} that describe the local effect of an infinitesimal change in concentration j on the rate of reaction i, i.e.

$$\varepsilon_{ij} = \left(\frac{\partial r_i}{\partial c_j}\right)^0 \cdot \frac{c_j^0}{r_i^0} \tag{3}$$

The ordinary differential equations (ODEs) were reformulated as differential algebraic equations (DAEs) to improve both the performance and stability of the numerical integrations, i.e. the conservation relations were solved algebraically. The DAE system was simulated with the linearly implicit differential algebraic solver «LIMEX» [15].

NETWORK PARAMETERIZATION FROM TIME-SERIES METABOLITE DATA

Parameterizing a kinetic model requires the specification of a reference steady state, i.e. J^0 and c^0, and the corresponding kinetic parameters, i.e. E_c^0. Each rate equation is assumed to be dependent on its substrate and product levels. In some instances additional effectors are taken into account;

The unknown elasticity coefficients and reference concentrations are identified by minimizing the differences between *in silico* model simulations and *in vivo* measurement data: The variance-weighted sum of squared residuals χ^2 between experimentally observed and simulated metabolite data, c^m and c^s, is minimized according to

$$\min_{E_c^0, c^0} \chi^2\left(E_c^0, c^0\right) = (c^s - c^m)^T \cdot \Sigma_m^{-1} \cdot (c^s - c^m) \tag{4}$$

Mauch, K. *et al.*

in which Σ_m is a diagonal matrix containing the measurement variances. An evolution strategy is applied for parameter fine-tuning that includes a self adapting mutation operator [16, 17].

The time-series metabolite data collected from HepG2 cell in the stimulus-response experiment were applied to parameterize the dynamic network model. Altogether, 174 scaled elasticities had to be estimated. Furthermore, 42 reference intermediate levels had to be identified (42 balanced compounds + 3 conserved moieties). The corresponding experimental data were available for 30 out of these. This means that in total 216 unknown parameters had to be specified in order to run a simulation. To enable a thorough exploration of the search space, the optimization runs were restarted after 100,000 evaluations of equation (4) using the best parameters currently available as starting values in the following iteration. Altogether, more than six million simulation runs were performed. On average, one simulation run took 0.2 seconds (Intel® Core2™ Quad CPU, 2.66 GHz, 4 GB RAM).

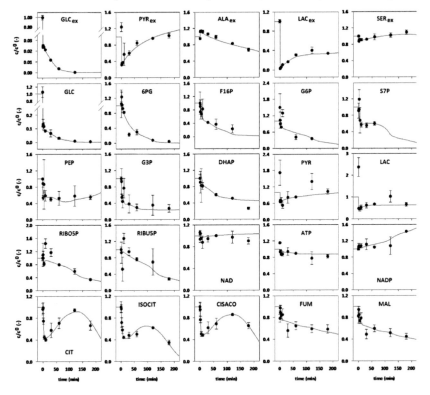

Figure 3. Experimentally determined (red circles) and simulated (blue solid lines) extra-/intracellular metabolite dynamics. The subscripts 'in' and 'ex' denote intracellular and extracellular metabolites, respectively. All concentrations were normalized with respect to the reference concentration directly before the stimulus. The error bars indicate standard deviations of the experimental data.

A total of 25 metabolite time courses were experimentally determined, of which 5 corresponded to extracellular metabolites and 20 to intracellular metabolites. The experimental data and the corresponding model simulations are summarized in Figure 3.

In vivo and *in silico* data were normalized with respect to the estimated reference values. It is worth noting that the perturbation triggered significant changes in the metabolite levels, and these changes provided important information about the underlying network dynamics. In general, the *in silico* metabolite dynamics were in accordance with the experimental data.

IN VIVO AND IN SILICO METABOLITE DYNAMICS

After exchanging the glucose-containing culture medium with the glucose-free medium, the extracellular glucose level dropped drastically. The remaining extracellular glucose was consumed by the cells within a period of 120 min. The extracellular pyruvate and lactate levels also dropped considerably because of the medium exchange, but started to accumulate again. At the end of the experiment, i. e. after 180 min the pyruvate values were even slightly higher than the estimated initial level. Lactate did not reach 50% of its initial value, which was the result of a decreasing lactate secretion rate. The initial efflux rate was twenty times higher for lactate than for pyruvate. This means that, in absolute terms, still more lactate than pyruvate was produced during the experiment. Extracellular alanine was consumed throughout the experiment, while extracellular serine accumulated. It is worth noting that besides the lack of glucose, the system was also perturbed as a result of the changes occurring in the extracellular pyruvate, lactate, alanine, and serine levels.

In accordance with the extracellular glucose levels, the intracellular glucose pool also decreased steeply. HepG2 cells have high GLUT2 transporter activities [18]. The GLUT2 transporter, which has a large K_m value, facilitates the diffusion of glucose into or out of the cells [19]. It can therefore be assumed that the steep decrease in the intracellular glucose pool was the result of the diffusion of intracellular glucose into the extracellular space. Consistently, the model simulations showed that the flux of glucose uptake was inversed immediately after the stimulus occurred. The intracellular glucose concentration further decreased and eventually converged to zero. With the exception of both phosphoenolpyruvate and pyruvate all other glycolytic metabolite levels decreased sharply immediately after the stimulus and continued to gradually decrease thereafter.

In the first 10 min of the experiment, the model simulations showed decreasing ribose-5-phosphate and ribulose-5-phosphate levels, followed by an increase, and then another decrease. Some discrepancy between the initial experimental data points and the simulations was observed for both metabolites, which could be an indication of a damped oscillation with rather high amplitude.

The first TCA cycle intermediate pools, i.e. citrate, cis-aconitate, and isocitrate, exhibited oscillatory dynamics. This was also found in the model simulations. It is interesting to note that three pairs of conjugate-complex eigenvalues were observed for the Jacobian matrix, which suggests that the system is capable of damped oscillations. The model simulations showed a non-oscillating decrease of fumarate and malate. There was some discrepancy between the simulated time courses for fumarate and malate and the experimentally observed concentrations after 30 min. However, the corresponding standard deviations were large.

The time courses of the experimentally determined cofactors NAD, ATP, and NADP only deviated slightly from their initial values. This means that despite the substantial changes in the metabolite levels in the central carbon metabolism, the homeostatic regulatory machinery of the hepatoma cells only allowed for small changes among the highly linked cofactors: ATP decreased slightly but remained at above 80% of its steady state concentration, the NAD level increased only marginally, NADP increased a little more, reaching 143% of its initial value. In contrast to these observations, distinct cofactor dynamics have been observed in similar stimulus response experiments in prokaryotes and yeast [20 – 22].

PERSPECTIVES

This contribution illustrates a framework for reconstructing and verifying kinetic metabolic network models in a semi-automated way. Since biochemical datasets like, for example, transcriptomic, proteomic, and metabolic data become increasingly available, the need as well as the potential for building predictive kinetic network models are gaining momentum. From a technical point of view, today's supercomputers and cloud computing environments allow for processing high-throughput experimental data in shorter time and at decreasing costs, thus enabling engineers and biochemists to use predictive kinetic network models as standard tool for decision making in areas like drug development and toxicity tests for the first time. From a more conceptual point of view, though, the application of large-scale predictive network models still faces a number of challenges. Fields such as (i) the coupling of signal transduction to metabolism, (ii) the systems-oriented breakdown of large-scale models into modules, and (iii) the integration of several modelling layers into concise representations of the biochemistry of whole organs and organisms are still not fully developed. Although enzyme kinetics form the building blocks of cellular dynamics, detailed kinetic information on many enzymes will probably remain rare and incomplete in the next decade. It is therefore anticipated that many questions at the systems-level will be answered through applying top-down/statistical approaches rather than through relying on bottom-up approaches.

ACKNOWLEDGEMENTS

Parts of this work were funded by HepatoSys (network detoxification), a systems biology funding initiative of the German Federal Ministry of Education and Research, and the Robert Bosch Foundation (Stuttgart, Germany). The excellent technical assistance of Sonja Seefried (Dr. Margarete Fischer-Bosch Institute of Clinical Pharmacology), Anja Niebel, and Gabriele Vacun (Institute of Biochemical Engineering, University of Stuttgart) is gratefully acknowledged. We also thank Jan G. Hengstler (Leibniz Research Centre for Working Environment and Human Factors at the University of Dortmund, Germany) for providing us with the HepG2 cell line.

REFERENCES

[1]   Hofmann, U. *et al.* (2008) Identification of metabolic fluxes in hepatic cells from transient 13C-labeling experiments: Part I. Experimental observations. *Biotechnol. Bioeng.* **100**:344 – 54.
        doi: http://dx.doi.org/10.1002/bit.21747.

[2]   Maier, K. *et al.* (2009) Quantification of statin effects on hepatic cholesterol synthesis by transient (13)C-flux analysis. *Metab. Eng.* **11**:292 – 309.
        doi: http://dx.doi.org/10.1016/j.ymben.2009.06.001.

[3]   Maier, K. *et al.* (2008) Identification of metabolic fluxes in hepatic cells from transient 13C-labeling experiments: Part II. Flux estimation. *Biotechnol. Bioeng.* **100**:355 – 70.
        doi: http://dx.doi.org/10.1002/bit.21746.

[4]   Hinkle, P.C. (2005) P/O ratios of mitochondrial oxidative phosphorylation. *Biochim. Biophys. Acta* **1706**:1 – 11, Jan 7 2005.
        doi: http://dx.doi.org/10.1016/j.bbabio.2004.09.004.

[5]   Schomburg, I. *et al.* (2002) BRENDA, enzyme data and metabolic information. *Nucleic Acids Res.* **30**:47 – 9.
        doi: http://dx.doi.org/10.1093/nar/30.1.47.

[6]   Schuster, S. and Schuster, R. (1991) Detecting Strictly Detailed Balanced Subnetworks in Open Chemical-Reaction Networks. *Journal of Mathematical Chemistry* **6**:17 – 40.
        doi: http://dx.doi.org/10.1007/BF01192571.

[7]   Visser, D. and Heijnen, J.J. (2003) Dynamic simulation and metabolic re-design of a branched pathway using linlog kinetics. *Metab. Eng.* **5**:164 – 76.
        doi: http://dx.doi.org/10.1016/S1096-7176(03)00025-9.

[8]   Visser, D. *et al.* (2004) Optimal re-design of primary metabolism in *Escherichia coli* using linlog kinetics. *Metab. Eng.* **6**:378–90.
        doi: http://dx.doi.org/10.1016/j.ymben.2004.07.001.

[9]   Westerhoff, H.V. and van Dam, K. (1987) Thermodynamics and Control of Biological Free-Energy Transduction. Elsevier, Amsterdam.

[10]   Heijnen, J.J. (2005) Approximative kinetic formats used in metabolic network modeling. *Biotechnol. Bioeng.* **91**:534–45.
        doi: http://dx.doi.org/10.1002/bit.20558.

[11]   Nikerel, I.E. *et al.* (2006) A method for estimation of elasticities in metabolic networks using steady state and dynamic metabolomics data and linlog kinetics. *BMC Bioinformatics* **7**: 540.
        doi: http://dx.doi.org/10.1186/1471-2105-7-540.

[12]   Hadlich, F., Noack, S., Wiechert, W. (2008) Translating biochemical network models between different kinetic formats. *Metab. Eng.* **11**(2):87–100.
        doi: http://dx.doi.org/10.1016/j.ymben.2008.10.002.

[13]   Hold, C. and Panke, S. (2009) Towards the engineering of *in vitro* systems. *J. R. Soc. Interface* **6**:507–521.
        doi: http://dx.doi.org/10.1098/rsif.2009.0110.focus.

[14]   Bulik, S. *et al.* (2009) Kinetic hybrid models composed of mechanistic and simplified enzymatic rate laws – a promising method for speeding up the kinetic modelling of complex metabolic networks. *FEBS J.* **276**:410–424.
        doi: http://dx.doi.org/10.1111/j.1742-4658.2008.06784.x.

[15]   Deuflard, P. *et al.* (1987) One Step and Extrapolation Methods for Differential-Algebraic Systems. *Numer. Math.* **51**:501–516.
        doi: http://dx.doi.org/10.1007/BF01400352.

[16]   Hansen, N. and Ostermeier, A. (2001) Completely derandomized self-adaption in evolutionary strategies. *Evolutionary Computation* **9**:159–195.
        doi: http://dx.doi.org/10.1162/106365601750190398.

[17]   Streichert, F. and Ulmer, H. (2005) JavaEvA: a Java based framework for Evolutionary Algorithms. *Technical Report* WSI-2005–06, Wilhelm-Schickard-Institut für Informatik (WSI), Center for Bioinformatics Tübingen (ZBIT), Eberhard-Karls-University Tübingen, Germany.

[18]   Wu, C.H. *et al.* (2009) *In vitro* and *in vivo* study of phloretin-induced apoptosis in human liver cancer cells involving inhibition of type ii glucose transporter. *Int. J. Cancer* **124**:2210–2219.

[19]   Thorens, B. (1996) Glucose transporters in the regulation of intestinal, renal, and liver glucose fluxes. *Am. J. Physiol.* **270**:541 – 553.

[20]   Chassagnole, C., Noisommit-Rizzi, N., Schmid, J.W., Mauch, K., Reuss, M. (2002) Dynamic modeling of the central carbon metabolism of *Escherichia coli. Biotechnol. Bioeng.* **79**(1):53-73.
         doi: 10.1002/bit.10288.

[21]   Magnus, J.B. *et al.* (2006) Monitoring and modeling of the reaction dynamics in the valine/leucine synthesis pathway in *Corynebacterium glutamicum. Biotechnology Progress* **22**:1071-1083.

[22]   Theobald, U., Mailinger, W., Baltes, M., Rizzi, M., Reuss, M. (1997) *In vivo* analysis of metabolic dynamics in *Saccharomyces cerevisiae: I. Experimental observations. Biotechnol. Bioeng.* **55**(2):305-316.
         doi: http://dx.doi.org/10.1002/(SICI)1097-0290(19970720)55:2<305::AID-BIT8>3.0.CO;2-M.

[23]   Veiga-da-Cunha, M. and Van Schaftingen, E. (2002) Identification of fructose 6-phosphate- and fructose 1-phosphate-binding residues in the regulatory protein of glucokinase. *J. Biol. Chem.* **277**:8466 – 73.
         doi: http://dx.doi.org/10.1074/jbc.M105984200.

[24]   Tsuboi, K.K. *et al.* (1971) Phosphoglucose isomerase from human erythrocyte. Preparation and properties. *J Biol. Chem.* **246**:7586 – 94.

[25]   Berg, J. *et al.* (2002) Biochemistry. Fifth Edition, International Version, New York, W.H. Freeman & Co.

[26]   Bloxham, D.P. and Lardy, H.A. (1973) Phosphofructokinase. In *The Enzymes.*Vol. 8, P. D. Boye (Ed.), 3rd ed. New York, Academic Press, pp. 239 – 278.

[27]   Bais, R. *et al.* (1985) The purification and properties of human liver ketohexokinase. A role for ketohexokinase and fructose-bisphosphate aldolase in the metabolic production of oxalate from xylitol. *Biochem. J.* **230**:53 – 60.

[28]   Gracy, R.W. (1975) Triosephosphate isomerase from human erythrocytes. *Methods Enzymol.* **41**:442 – 7.

[29]   Gregus, Z. and Nemeti, B. (2005) The glycolytic enzyme glyceraldehyde-3-phosphate dehydrogenase works as an arsenate reductase in human red blood cells and rat liver cytosol. *Toxicol. Sci.* **85**:859 – 69.
         doi: http://dx.doi.org/10.1093/toxsci/kfi158.

[30]   Scopes, R.K. (1973) 3-Phosphoglycerate kinase. In *The Enzymes.* Vol. 8, P. D. Boyer (Ed.), 3 ed New York, Academic Press, pp. 335 – 351.

[31]   Staal, G.E.J. *et al.* (1975) Human erythrocyte pyruvate kinase. *Methods Enzymol.* **42C**:182 – 186.
doi: http://dx.doi.org/10.1016/0076-6879(75)42113-9.

[32]   Dombrauckas, J.D. *et al.* (2005) Structural basis for tumor pyruvate kinase M2 allosteric regulation and catalysis. *Biochemistry* **44**:9417 – 29.
doi: http://dx.doi.org/10.1021/bi0474923.

[33]   Kahn, A. and Marie, J. (1982) Pyruvate kinases from human erythrocytes and liver. Methods Enzymol. **90** Pt E: 131 – 40.

[34]   Cho, S.W. and Joshi, J.G. (1990) Characterization of glucose-6-phosphate dehydrogenase isozymes from human and pig brain. *Neuroscience* **38**:819 – 28.
doi: http://dx.doi.org/10.1016/0306-4522(90)90074-E.

[35]   Fazi, A. *et al.* (1990) Purification and partial characterization of the phosphogluco-mutase isozymes from human placenta. *Prep. Biochem.* **20**:219 – 240.
doi: http://dx.doi.org/10.1080/00327489008050198.

[36]   Stryer, L. (1995) Biochemistry. W. H. Freeman and Company, Vol. 4.

[37]   Gibson, G.E. *et al.* (2000) The alpha-ketoglutarate dehydrogenase complex in neuro-degeneration. *Neurochem. Int.***36**:97 – 112.
doi: http://dx.doi.org/10.1016/S0197-0186(99)00114-X.

[38]   Ogata, H. *et al.* (1999) KEGG: Kyoto Encyclopedia of Genes and Genomes. *Nucleic Acids Res.* **27**:29 – 34.

[39]   Kanehisa, M. and Goto, S. (2000) KEGG: kyoto encyclopedia of genes and genomes. *Nucleic Acids Res.* **28**:27 – 30.
doi: http://dx.doi.org/10.1093/nar/28.1.27.

[40]   Soundar, S. *et al.* (2003) Evaluation by mutagenesis of the importance of 3 arginines in alpha, beta, and gamma subunits of human NAD-dependent isocitrate dehydro-genase. *J. Biol. Chem.* **278**:52146 – 53.
doi: http://dx.doi.org/10.1074/jbc.M306178200.

[41]   Lazo, P.A. and Sols, A. (1980) Pyruvate dehydrogenase complex of ascites tumour. Activation by AMP and other properties of potential significance in metabolic regulation. *Biochem. J.* **190**:705 – 10.

Beilstein-Institut

137

Experimental Standard Conditions of Enzyme Characterizations,
September 13th – 16th, 2009, Rüdesheim/Rhein, Germany

Interferon-γ Stimulated STAT1 Signalling: from Experimental Data to a Predictive Mathematical Model

Katja Rateitschak[1,*], Robert Jaster[2], Olaf Wolkenhauer[1]

[1]Systems Biology & Bioinformatics, University of Rostock, 18051 Rostock, Germany

[2]Department of Medicine II, Division of Gastroenterology, Medical Faculty, University of Rostock, 18057 Rostock, Germany

E-Mail: *katja.rateitschak@uni-rostock.de

Received: 15th March 2010 / Published: 14th September 2010

Systems Biology

Systems biology aims to understand the complex dynamics of biochemical reaction networks by an interdisciplinary approach of mathematical modelling and quantitative cell biology [1 – 6]. Stimulating an intracellular signalling pathway by an extracellular ligand leads to temporal changes of intracellular protein concentrations. These temporal changes are caused by nonlinear processes, including protein complex formation, enzyme catalyzed reactions and feedback regulation. Thus simple measurements of experimental time courses for protein concentrations or measurements of reaction rate constants alone are not sufficient to understand the dynamics of a biochemical network. Mathematical models, describing the temporal response of network in response to systematic perturbations of other components are needed to unravel the nonlinear dynamics of biochemical networks [7, 8].

The properties of mathematical models can be investigated through formal analysis or numerical computer simulations. This approach allows us to explore relationships between structure and function of pathways by analysing mathematical models with methods from dynamical systems theory [9 – 11]. A model can only be as good as the data that were used to build it and hence the generation of quantitative, sufficiently rich time series is a major priority for systems biology. Towards this end, the development of error reduction strategies for standardised quantitative data [12], optimisation methods to estimate reaction constants from experimental data [13 – 16], and design of experiments on the basis of model predictions has received attention [17, 18].

An important application of medical systems biology is to understand disease mechanisms and the response of cells to drugs in order to support the development of therapies. Computer simulations enable quantitative predictions of sRNAi experiments and the design of stimulus-profiles leading to a desired temporal response of a pathway target. But only recently, feedback to the wet lab in the form of quantitative predictions of pathway dynamics and supporting the design of experiments is taking place as we demonstrate in our work.

WORKFLOW IN SYSTEMS BIOLOGY

A fruitful systems biology project requires a close interaction between scientists from cell biology and those with experience in mathematical modelling. A typical workflow in systems biology follows a bottom-up strategy. It starts with a single signalling pathway, composed of few core proteins and thus allows a detailed analysis of its dynamical properties. The initial network will be refined and extended by an iterative cycle of quantitative cell biology, mathematical modelling, parameter estimation, model predictions and experimental validation of model predictions. A systems biology workflow is summarized in Figure 1. The objectives of a specific systems biology project determine the type of the experimental data needed and both determine the mathematical modelling approach. In the following, we describe the steps of the workflow in more detail. For each step a variety of approaches are possible but we restrict to selected examples.

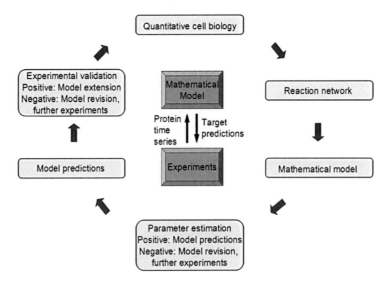

Figure 1. Workflow in systems biology.

Step 1: Quantitative cell biology

An appropriate approach to study the influence of a signalling pathway on cell proliferation is the *in vitro* stimulation of a pathway in a standardized cellular background. If cellular decisions are relevant (as in differentiation or apoptosis), then also flow cytometry data are needed which allow a sorting of cells with qualitatively different responses. In the above experiments temporal concentration changes are measured as an average over many cells. Protein concentration changes are quantified by immunoblots or by ELISA tests and gene expression changes of feedback target genes are quantified by Real-time PCR.

Step 2: Reaction network and structure of the mathematical model

The biochemical reaction network is translated into a nonlinear differential equation model, describing temporal changes of protein and mRNA concentrations. Ordinary differential equations (ODE) are an appropriate modelling approach if the experiments involve large numbers of proteins and mRNAs such that concentrations smoothly change over time. This is the case in the above introduced experiments. The mathematical models can include:

1. Receptor activation

2. Enzyme catalyzed reactions, for example phosphorylations of proteins

3. Protein complex formation

4. Intracellular transport and shuttling of proteins between subcellular compartments

5. Target gene expression and feedback regulation by expressed target genes.

Slow processes, for example gene transcription, should be appropriately modelled by delay differential equations [19]. In general, one applies mass action kinetics. Under certain assumptions models can be simplified, for example Michaelis-Menten kinetics.

Step 3: Parameter estimation

The parameters of a model, including reaction constants, delay times and total protein concentrations, are the remaining unknown components of the model. They can be estimated from ELISA time series, immunoblotting, microscopy and real-time PCR by global optimization approaches. Very fast scatter search algorithms with alternating global and local search have been developed for systems biology applications [16].

Step 4: Model validation and model-based predictions

Once model structure and parameter values are determined, the properties of a mathematical model are studied by formal analysis and computer simulations. However, a mathematical model is of benefit to the laboratory work only if it can predict the outcome of experiments, which have not been used for model calibration. The design of refined experiments by model predictions will improve our understanding of biochemical networks. An important application of model driven experimental design is to achieve a deeper pathophysiological understanding of biochemical processes and to get knowledge about steps in the network which are difficult to measure. Mathematical models can be applied to predict an optimal range of stimulus concentrations, an optimal combination of different stimuli, as well as the kind and duration of stimulation to receive a desired temporal profile of the output signal.

If model predictions cannot be experimentally validated then one has to go back to *Step 2: Reaction network and structure of mathematical model.*

MODELLING STRATEGY

A useful approach starts with a model which includes only those proteins for which experimental data are available, and interactions indirectly indicated by the time courses to be necessary in the biochemical process. An example is an oscillatory time series, but so far no negative feedback protein is known in the respective network. Different hypotheses about interactions, for example different feedback mechanisms through expressed target genes, or different transport mechanisms will be tested by different model structures.

If the mathematical model does not fit the experimental data then the following two reasons are possible: Important biochemical reactions are missing or reactions are not appropriately mathematically described. In the first case, a feedback loop or crosstalk with another signalling pathway may have not been considered and in the second case a delay differential equation could be more useful instead of an ordinary differential equation due to a time lag between the dynamics of the respective variables. Thus, one has to go back to *Step 2: Reaction network and structure of mathematical model.* Information about missing elements or inappropriate assumptions can be obtained also from the bad fit, as we will demonstrate in our example of the STAT1 signalling pathway.

An alternative strategy to the approach described above includes all known interactions of a biochemical network in a mathematical model. For example, given that there are 20 key proteins known, this leads to about 50 parameters in the model. Assuming that time series from only three proteins are available for parameter estimation. It is much easier to fit a large model with about 50 parameters to three experimental time series. Even many parameter sets will lead to a very good fit. Such a model is called unidentifiable. Predictions of an unidentifiable model can depend on the chosen parameter set of the best fits [18].

In the following sections we will apply our systems biology workflow to study the interferon-γ (IFNγ) stimulated STAT1 signalling pathway in pancreatic stellate cells (PSC); a fibroblast-like cell type in the pancreas.

INTERFERON-γ STIMULATED STAT1 SIGNALLING PATHWAY IN PSC

Cells sense their environment through receptors. Signal transduction is the transfer of this information through a cell and between cells. Frequently, the immediate targets of signal transduction are transcription factors, which regulate the expression of specific target genes. The corresponding gene products mediate an appropriate physiological response to environmental changes. The family of STAT proteins (signal transducer and activator of transcription) exerts a dual role as signalling protein and transcription factor [20]. They transfer information directly from the receptor to the target genes. The STAT proteins can be activated by different receptors, and STAT signal transduction can influence different cellular functions, like proliferation, differentiation and immune response [20].

We studied the IFNγ stimulated STAT1 signalling pathway in PSC by a systems biology approach. STAT1 plays a key role in the mediation of antifibrotic effects of IFNγ effects in PSC [21, 22]. A detailed understanding of regulatory mechanisms controlling STAT1 activity in PSC may contribute to the development of a targeted antifibrotic therapy. In this work we present a summary of our results in [23] and few additional simulations. Details about the experimental protocol, quantified experimental data, mathematical model and simulation results are presented in [23].

MODEL OF THE IFNγ STIMULATED STAT1 SIGNALLING PATHWAY

Immunoblot time series for STAT1 activation, immunofluorescence time series for its nuclear translocation and RT-PCR time series for the expression of the target gene *SOCS1* measured in immortalised PSC formed the starting point for our systems biology project.

Based on the experimental data, we first established a reduced reaction network of the IFNγ-stimulated STAT1 signalling pathway (Fig. 2).

The reaction network includes phosphorylation of unphosphorylated STAT1 (STAT1U) followed by rapid homodimer formation (STAT1D). STAT1D translocate into the nucleus and induce the transcription of specific target genes. When STAT1D is not bound to the DNA, dimers may dissociate, followed by protein dephosphorylation and nuclear export of the resulting STAT1U [24]. In the network it is also included that STAT1U can shuttle into the nucleus, where it induces the transcription of target genes [25]. The network considers *SOCS1* and *STAT1* itself as target genes of IFNγ-activated signalling through STAT1. The annotation *time delay* in Figure 2 refers to temporal differences between IFNγ action at the receptor level (including STAT1 activation) and consecutive steps, such as expression of target genes.

Rateitschak, K. *et al.*

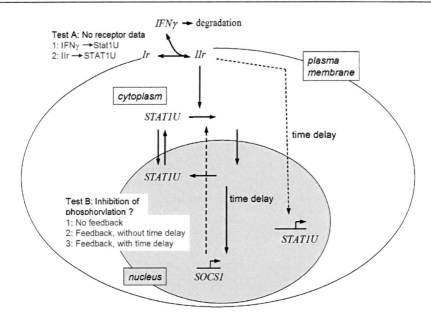

Figure 2. Reaction network of the IFNγ stimulated STAT1 signalling pathway. Modified reprint from [23] with permission from Elsevier.

So far, we have no experimental data for receptor activation. Therefore, we tested two models (Test A):

1. IFNγ mediates the phosphorylation of STAT1U

2. IFNγ reversibly binds to the interferon receptor (Ir) leading to receptor activation. The active complex is denoted by IIr.

SOCS1 is a potential negative feed-back regulator inhibiting the phosphorylation of STAT1U [26, 27]. Our data show that levels of phosphorylated STAT1 (STAT1P) did not decrease for IFNγ = 100 ng/ml (Fig. 3B, right) despite induction of *SOCS1* expression. This observation is compatible with two scenarios:

(1) The SOCS1 concentration is low compared to the number of receptors. In this case, the effect of SOCS1 on the reduction of STAT1P phosphorylation would be negligible.

(2) The negative feedback induced by SOCS1 could be effective at late times (≥ 180 min) only, where it could reduce the slope of the late increase of STAT1P.

To further study the role of SOCS1, we tested the following three hypotheses by mathematical modelling (Test B):

1. No feedback inhibition by SOCS 1

2. Feedback inhibition without time delay

3. Feedback inhibition with time delay

The reaction network was translated into a system of ordinary differential equations (ODE), which describe temporal changes of the network components as a function of interactions and transport processes. The delayed processes of increased STAT1U expression, *SOCS1* transcription and hypothetical negative feedback by SOCS 1 were described by a distributed time delay. All parameters were estimated on the basis of time series for STAT1, STAT1P and *SOCS1* mRNA by global optimisation using the fSSmb algorithm [14].

MODEL SIMULATIONS AND QUANTITATIVE PREDICTIONS

Figure 3 shows a comparison between the experimental time series and the simulation results of the mathematical models (using optimised parameters) based on different hypotheses.

First, we compared the parameter estimates of the two models of Test A in Figure 3A: Computer simulations of the model without receptor lead to good fit for all observables (data not shown) except of a poor fit for STAT1. The poor fit indicates that the reaction rate describing STAT1 transcription is too small. Therefore we included reversible receptor activation in a revised model to get a longer persisting signal for STAT1 transcription. The parameter estimates for this model lead to a good fit for all observables including STAT1U as shown in Figure 3A.

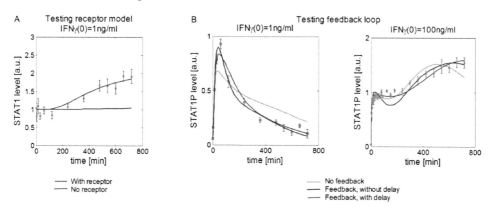

Figure 3. Comparison between (A) experimental time series and (B) computer simulations of the different models with estimated parameters. Modified reprint from [23] with permission from Elsevier.

Next, we compared the parameter estimates of the three models of Test B in Figure 3B: Our simulation results show that the model with the delayed negative feedback by Socs1 leads to the best fit of the experimental data for STAT1P compared with the models without a feedback and a non-delayed feedback, respectively.

Note, that the differences between the "best fit" model and the other two models of test B are small. A SOCS1 siRNA knockdown experiment can reveal whether SOCS1 negatively regulates STAT1 phosphorylation. If the outcome of this experiment will be negative then the identification of other interactions, for example other feedback loops or crosstalk with other signalling pathways can help to improve the model.

To validate the mathematical model with the IFNγ receptor and with the delayed negative feedback by SOCS1, we tested its ability to predict the results of experiments not previously used for model calibration. We studied by computer simulations the influence of the stimulation scenario on the network dynamics. The simulation results for a stimulation with a single dose of IFNγ (1 ng/ml) and for multiple stimulations with smaller doses (4 x 0.25 ng/ml; applied at intervals of three hours) are shown in Figure 4. The model predictions have been experimentally validated. The experimental data are also shown in Figure 4. Even the complex staircase structure in the experimental time series of STAT1P (split-dose mode) was correctly indicated by the mathematical model. Furthermore predicted and measured slopes of increases were similar for most time points.

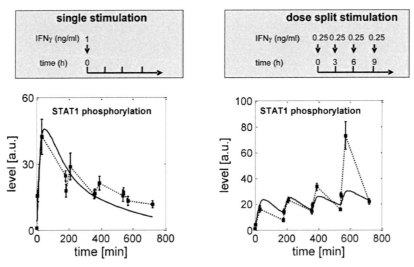

Figure 4. Model predictions and experimental validation. Reprint from [23] with permission from Elsevier.

ACKNOWLEDGEMENTS

This work was supported by grants from the Bundesministerium für Bildung und Forschung through the FORSYS partner programme (grant number 0315255 to K. R.) and the Deutsche Forschungsgemeinschaft (to R. J.). O.W. acknowledges support from the Helmholtz Society as part of the systems biology network.

REFERENCES

[1]   Kitano, H. (2002) Computational systems biology. *Nature* **420**:206 – 210.
        doi: http://dx.doi.org/10.1038/nature01254.

[2]   Alon, U. (2007) Am introduction to systems biology. Chapman & Hall/CRC.

[3]   Palsson, B. (2006) Systems Biology: Properties of Reconstructed Networks. Cambridge University Press.

[4]   Kirschner, M. (2005) The meaning of systems biology. *Cell* **121**:503 – 504.
        doi: http://dx.doi.org/10.1016/j.cell.2005.05.005.

[5]   Systems Biology – Nobel Symposium (2009) *FEBS Letters* **583**:3881 – 4030.

[6]   Ferrell, J.E. (2009) Q&A: Systems Biology. *Journal of Biology* **8**:2.
        doi: http://dx.doi.org/10.1186/jbiol107.

[7]   Kholodenko, B.N., Kiyatkin, A.,   Bruggeman, F.J., Sontag E., Westerhoff, H.V., Hoek, J.B. (2002) Untangling the wires: A strategy to trace functional interactions in signaling and gene networks *Proc. Natl. Acad. Sci. U.S.A.* **99**:12841 – 12846.
        doi: http://dx.doi.org/10.1073/pnas.192442699.

[8]   Santos, S.D.M., Verveer, P.J., Bastiaens, P.I.H. (2007) Growth factor-induced MAPK network topology shapes Erk response determining PC-12 cell fate. *Nat. Cell Biol.* **9**:324 – 336.
        doi: http://dx.doi.org/10.1038/ncb1543.

[9]   Klipp, E. *et al.* (2005) Systems Biology in Practice. Wiley-VCH Weinheim.
        doi: http://dx.doi.org/10.1002/3527603603.

[10]   Perko, L. (2000) Differential Equations and Dynamical Systems. Springer New York.

[11]   Strogatz, S.H. (2001) Nonlinear Dynamics and Chaos: With Applications to Physics, Biology,Chemistry and Engineering. Boulder, CO: Westview Press.

[12] Schilling, M., Maiwald, T., Bohl, S. Kollmann, M., Kreutz, C., Zimmer, J., Kling-mueller, U. (2005) Computational processing and error reduction strategies for stan-dardized quantitative data in biological networks. *FEBS Journal* **272**:6400 – 6412.
doi: http://dx.doi.org/10.1111/j.1742-4658.2005.05037.x.

[13] Moles, C.G., Mendes, P., Banga, J.R. (2003) Parameter estimation in biochemical pathways: a comparison of global optimization methods. *Genome Research* **13**:2467 – 2474.
doi: http://dx.doi.org/10.1101/gr.1262503.

[14] Kutalik, Z., Cho, K.H., Wolkenhauer O. (2004) Optimal Sampling Time Selection for Parameter Estimation in Signal Transduction Pathway Modelling. *BioSystems* **75**:43 – 55.
doi: http://dx.doi.org/10.1016/j.biosystems.2004.03.007.

[15] Swameye, I., Mueller, T.G., Timmer, J., Sandra, O., Klingmueller, U. (2003) Identi-fication of nucleoplasmic cycling as remote sensor in cellular signaling by database modelling. *Proc. Natl. Acad. Sci. U.S.A.* **100**:1028 – 1033.
doi: http://dx.doi.org/10.1073/pnas.0237333100.

[16] Rodriguez-Fernandez, M., Egea, J.A., Banga, J.R. (2006) Novel metaheuristic for parameter estimation in nonlinear dynamic biological systems. *BMC Bioinformatics* **7**:483.
doi: http://dx.doi.org/10.1186/1471-2105-7-483.

[17] Kreutz, C.,Timmer, J. (2009) Systems biology: experimental design. *FEBS Journal* **276**:923 – 942.
doi: http://dx.doi.org/10.1111/j.1742-4658.2008.06843.x.

[18] Raue, A., Kreutz, C., Maiwald, T., Bachmann, J., Schilling, M., Klingmüller U., Timmer, J. (2009) Structural and practical identifiability analysis of partially ob-served dynamical models by exploiting the profile likelihood. *Bioinformatics* **25**:1923 – 1929.
doi: http://dx.doi.org/10.1093/bioinformatics/btp358.

[19] Rateitschak, K., Wolkenhauer, O. (2007) Intracellular delay limits cyclic changes in gene expression. *Mathematical Biosciences* **205**:163 – 179.
doi: http://dx.doi.org/10.1016/j.mbs.2006.08.010.

[20] Levy, D. E., Darnell Jr.,J.E. (2002) Stats: Transcriptional control and biological impact. *Nat. Rev. Mol. Cell. Biol.* **3**:651 – 662.
doi: http://dx.doi.org/10.1038/nrm909.

[21] Baumert, J.T., Sparmann, G., Emmrich, J., Liebe, S., Jaster, R. (2006) Inhibitory effects of interferons on pancreatic stellate cell activation. *World J. Gastroenterol.* **12**:896 – 901.

[22]   Fitzner, B., Brock, P., Nechutova, H., Glass, A., Karopka, T., Koczan, D., Thiesen, H.J., Sparmann, G., Emmrich, J., Liebe, S., Jaster, R. (2007) Inhibitory effects of interferon-gamma on activation of rat pancreatic stellate cells are mediated by STAT1 and involve down-regulation of CTGF expression. *Cellular Signalling* **19**:782 – 790.
doi: http://dx.doi.org/10.1016/j.cellsig.2006.10.002.

[23]   Rateitschak, K., Karger A., Fitzner B. Lange F., Wolkenhauer O. Jaster R. (2010) Mathematical modelling of interferon-γ signalling in pancreatic stellate cells reflects and predicts the dynamics of STAT1 pathway activity. *Cellular Signalling* **22**:97 – 105.
doi: http://dx.doi.org/10.1016/j.cellsig.2009.09.019.

[24]   Begitt, A., Meyer, T., van Rossum, M., Vinkemeier, U. (2000) Nucleocytoplasmic translocation of Stat1 is regulated by a leucine-rich export signal in the coiled-coil domain. *Proc. Natl. Acad. Sci. U.S.A.* **97**:10418 – 10423.
doi: http://dx.doi.org/10.1073/pnas.190318397.

[25]   Yang, J., Stark, G.R. (2008) Roles of unphosphorylated STATs in signaling. *Cell Research* **18**:443 – 451.
doi: http://dx.doi.org/10.1038/cr.2008.41.

[26]   Wormald, S., Zhang, J.G., Krebs, D.L., Mielke, L.A., Silver, J., Alexander, W.S., Speed, T.P., Nicola, N.A., Hilton, D.J. (2006) The comparative roles of suppressor of cytokine signalling-1 and -3 in the inhibition and desensitization of cytokine signalling. *J. Biol. Chem.* **281**:11135 – 11143.
doi: http://dx.doi.org/10.1074/jbc.M509595200.

[27]   Endo, T.A., Masuhara, M., Yokouchi, M., Suzuki, R., Sakamoto, H., Mitsui, K., Matsumoto, A., Tanimura, S. Ohtsubo, M., Misawa, H., Miyazaki, T., Leonor, N., Taniguchi, T. Fujita, T., Kanakura, Y., Komiya, S., Yoshimura A. (1997) A new protein containing an SH2 domain that inhibits JAK kinases. *Nature* **387**:921 – 924.
doi: http://dx.doi.org/10.1038/43213.

Using Generalised Supply-Demand Analysis to Identify Regulatory Metabolites

Johann M. Rohwer[1,*] and Jan-Hendrik S. Hofmeyr[1,2]

[1]Triple-J Group for Molecular Cell Physiology, Department of Biochemistry, and

[2]Centre for Studies in Complexity, Stellenbosch University, Private Bag X1, ZA-7602 Matieland, South Africa

E-Mail: *jr@sun.ac.za

Received: 22nd February 2010 / Published: 14th September 2010

Abstract

We present the framework of generalised supply-demand analysis of a kinetic model of a cellular system, which can be applied to networks of arbitrary complexity. By fixing the concentrations of each of the variable species in turn and varying them in a parameter scan, rate characteristics of supply-demand are constructed around each of these species. The shapes of the rate characteristic patterns and the magnitude of the flux-response coefficients of the supply and demand blocks, as compared to the elasticities of the enzymes that interact directly with the fixed metabolite, allow for identification of regulatory metabolites in the system. The analysis provides information not only on whether and where the system is functionally differentiated, but also on the degree to which of its species are homeostatically buffered. The novelty in our method lies in the fact it is unbiased, supplying an entry point for the further analysis and detailed characterisation of large models of cellular systems, in which the choice of metabolite around which to perform a supply-demand analysis is not always obvious. The method is exemplified with two kinetic models from the published scientific literature.

1 Introduction

The burgeoning field of systems biology (*e. g.* [1]) has developed out of the realisation that biological systems cannot be understood from reductionist characterisation of their components alone, but that their interactions have to be integrated in a "systems" framework. New models of diverse cellular pathways appear regularly and are curated and stored in online databases [2, 3]; such models provide powerful tools that are often more accessible to query and interrogation than experimental systems. Yet without proper frameworks of analysis, these models, albeit big and comprehensive, remain little more than collections of data.

The framework of supply-demand analysis (SDA), developed by Hofmeyr and Cornish-Bowden [4], has proved useful in studying the regulation of cellular pathways within the metaphor of an economy controlled by supply and demand. It has become a reference framework for analysing metabolic pathways by teaching scientists to look for flux control beyond the scope of what has traditionally been called the pathway, i. e. in the demand for its end-product, a view that has subsequently been corroborated by experimental data (*e. g.* [5]). While SDA can provide useful results, its application to large kinetic models of cellular pathways is, however, hampered by the problem that their complexity may preclude us from finding a "natural" subdivision of the system into supply and demand blocks. With this in mind, we have generalised SDA so that it can be applied to models of arbitrary size and complexity in a systematic, computer-driven way [6].

In the remainder of this paper we first summarise generalised supply-demand analysis as developed in [6] (Section 2), and subsequently apply it to two published kinetic models (Section 3).

2 Principles and Illustration of Generalised Supply-Demand Analysis

Ordinary supply-demand analysis (SDA) [4] considers the subdivision of a pathway around a central intermediate, with the block or blocks of reactions contributing to the production of the intermediate constituting the "supply", and those that contribute to its consumption, the "demand". The behaviour of the system around the steady-state point is assessed with a so-called combined rate characteristic [7], which depicts how the rates of supply and demand vary with changes in the concentration of the intermediate. The intersection of the supply and demand rate characteristics signifies the steady-state point. If the rate characteristic is drawn in double-logarithmic space, the elasticities [8] of supply and demand towards the intermediate can be read off directly as slopes of the tangents to the supply and demand curves at the steady-state point, enabling the calculation of control coefficients. One of the main tenets of SDA is that when one of the two blocks controls the flux, the other one determines the degree of homeostasis in the intermediate; such a system has been termed *functionally differentiated* [4].

In silico SDA with a kinetic model can be easily achieved by making the intermediate around which the rate characteristic is to be constructed, into a fixed (clamped) species of the model, thus turning it into a model parameter. This parameter is then varied over a wide range through a parameter scan. An implicit assumption of this approach is that the system has been or can readily be partitioned into supply and demand; however, when faced with the complexity of cellular pathways or of large models of such pathways, the choice of intermediate around which to perform the SDA is often far from obvious.

To address this shortcoming, we have generalised SDA in such a way that it can easily be performed on a kinetic models of any cellular system, large or small, without requiring prior knowledge of its regulatory structure [6]. Generalised supply-demand analysis (GSDA) works in the following way: *Each* of the variable intermediates is *clamped in turn* and thus made into a parameter of the system. Its concentration is then varied above and below the reference steady-state value in the original system through a parameter scan, and the fluxes through the supply and demand reactions that are directly connected to the intermediate are plotted on a log-log rate characteristic. Every flux that directly produces the intermediate is a separate supply flux, and likewise, each flux that directly consumes it is a separate demand flux. There will thus be as many rate characteristics as there are reactions that produce or consume the intermediate. It should be emphasised that this procedure is valid for arbitrary models and does not presuppose a subdivision of the system into supply and demand blocks.

GSDA yields as many combined rate characteristic graphs as there are variable species in the system. As will be shown below, the following important features about the regulation of the system can be identified from the shapes of the curves and associated elasticities and response coefficients:

1. Potential sites of regulation;

2. Regulatory metabolites;

3. The quantitative relative contribution of different routes of interaction from an intermediate to a supply or demand block;

4. Sites of functional differentiation where one of the supply or demand blocks predominantly controls the flux, and the other determines the degree of homeo-static buffering of the intermediate.

We next exemplify GSDA with a model of a linear 5-enzyme pathway containing a feedback loop.

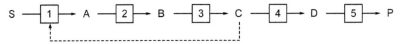

Figure 1. A five-enzyme linear pathway converting substrate S to product P. In one of the models, the first enzyme is allosterically inhibited by intermediate C (see main text). Reproduced with permission from [6].

2.1 GSDA of a Simple Linear Pathway

The simulations in this section use two variants of a kinetic model of the linear pathway in Figure 1 (detailed model descriptions and computational methods are provided in [6]).

Model I This is the base-line, undifferentiated version in which all five enzymes have identical kinetic parameters and are modelled with reversible Michaelis-Menten kinetics (with the exception of enzyme 5, which is modelled with irreversible Michaelis-Menten kinetics). There is no allosteric feedback from C to enzyme 1.

Model II In this model, enzyme 1 is inhibited allosterically by C and is modelled with reversible Hill kinetics [9]. The limiting rates of enzymes 2 and 3 have been increased so that they are close to equilibrium. Enzymes 4 and 5 together have almost complete control over the flux through the pathway.

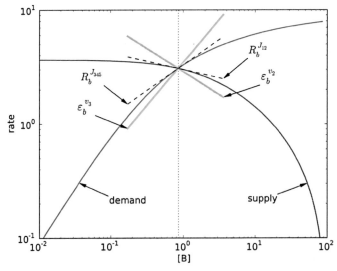

Figure 2. Supply-demand analysis around metabolite B for Model I. The concentration of B was clamped and varied to generate the supply and demand rate characteristics, as described in the text. The steady-state concentration is indicated by a vertical dotted line. The rate characteristics, response coefficients (tangents to the rate characteristics at the steady-state point) and elasticities of the supply and demand enzymes directly connected to B are labelled on the graph. Reproduced with permission from [6].

As explained above, a GSDA is performed by clamping each variable species of the model in turn and varying its concentration to generate the supply and demand rate characteristics. This yields graphs such as in Figure 2, which shows the GSDA around metabolite B in model I. To facilitate the interpretation of such graphs, this specific case is discussed in detail. The intersection of the log-log rate characteristics of supply and demand marks the steady-state point. The supply rate characteristic is drawn in blue and the demand rate characteristic in green. The slopes of the tangents to the rate characteristics (indicated by dashed lines on the graph) equal the flux-response coefficients (see e.g. [8]) of supply and demand towards B (in Figure 2, J_{12} signifies the flux through the supply block and J_{345} that through the demand block). These response coefficients quantify how sensitively the supply and demand fluxes respond towards changes in b, and are equivalent to "block-elasticities" [10] or co-response coefficients [11, 12] in the complete system where B is not clamped.

SDA as originally described [4] assumes that the only communication between supply and demand is via the linking intermediate. In this situation, the supply-demand block control coefficients of the complete pathway can be directly calculated from the supply and demand block elasticities, and the distribution of flux control is determined by the ratio of the block elasticities, while the magnitude of concentration control is determined by the difference $\varepsilon_b^{v_{345}} - \varepsilon_b^{v_{12}}$. Figure 2 also shows graphically the elasticities of the enzymes that produce or consume B. Elasticities are local properties of enzymes and quantify how sensitively an enzyme's local rate responds to changes in a substrate, product or effector. They are in fact apparent kinetic orders. In this case B is a product of v_2 and a substrate for v_3, so Figure 2 shows $\varepsilon_b^{v_2}$ (thick light blue line) and $\varepsilon_b^{v_3}$ (thick light green line).

The crux of GSDA now lies in the comparison of the values of the response coefficients with the elasticities of the enzymes that are directly connected to the clamped metabolite. In Figure 2, these values differ, i.e. $R_b^{J_{12}} \neq \varepsilon_b^{v_2}$ and $R_b^{J_{345}} \neq \varepsilon_b^{v_3}$. In other cases, they will be seen to agree. However, before comparing them in detail, first we have to present the GSDA of all metabolites for both models.

The graphs in Figure 3 present the results of the GSDA on models I and II. To avoid clutter, the graphs are not annotated but they follow the same convention as Figure 2. The only additional piece of information required is that of an allosteric modifier elasticity ($\varepsilon_c^{v_1}$ in Fig. 3b with C clamped, as only model II has the feedback loop). This is drawn in an orange line to set it apart from the supply and demand elasticities.

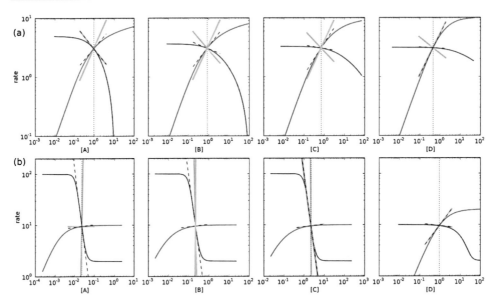

Figure 3. Generalised supply-demand analysis of the system depicted in Fig. 1. The concentrations of A–D were clamped in turn and varied to generate the supply and demand rate characteristics, as described in the text. The supply rate characteristic is drawn in blue, that for the demand in green. The steady-state concentration of the clamped metabolite is indicated by a vertical dotted line. The response coefficients of the supply and demand blocks are indicated by black dashed lines. The elasticities of the supply and demand enzymes for the clamped intermediate they are directly connected to are indicated by thick lines of the same colour as the rate characteristic. Model variants: (a) model I, (b) model II (see main text). In (b), the allosteric elasticity $\varepsilon_c^{v_1}$ is indicated by an orange thick line. Adapted from [6] with permission.

2.2 Interpretation of Generalised Supply-demand Analysis Graphs

The graphs in Figure 3 contain a wealth of information. As shown in [6], they can be interpreted on four levels, i.e. differences in the rate characteristic shapes as one proceeds from one metabolite to the next in the pathway, comparison of elasticity and response slopes, identification of points of functional differentiation and homeostasis, and finally, refined analysis through partial response coefficients.

DIFFERENCES IN RATE CHARACTERISTIC SHAPES

The first assessment criterion of GSDA merely looks at the general shapes of the supply and demand rate characteristics and is not yet concerned with elasticities and response coefficients. In model I (Fig. 3a) all enzymes have identical kinetics and the overall shapes of the rate characteristics are similar for metabolites A–D. In model II (Fig. 3b), however, the pattern for D is different from those for A–C (which are still similar). This means that the kinetic properties of enzyme 4 are such that site of regulation has been introduced into the system. In this specific case the reason is that enzyme 4 has been made insensitive to changes in the concentration of C ($\varepsilon_c^{v_4} \approx 0$). In general, such zero elasticities, whether towards substrate or product, induce a change in the rate characteristic shape because they shift the flux control to demand or supply, respectively. Overall, changes in the rate characteristic shapes thus pin-point potential sites of regulation.

COMPARISON OF ELASTICITIES AND RESPONSE COEFFICIENTS

GSDA can be extended to a second level by comparing the values of the elasticities and flux-response coefficients at the steady-state point for each metabolite. From the partitioned response property of control analysis [8],

$$R_p^J = \varepsilon_p^{v_i} \times C_{v_i}^J \tag{1}$$

it follows that $R_p^J = \varepsilon_p^{v_i}$ if $C_{v_i}^J = 1$. This means that the enzyme on which the intermediate acts directly must have full control over its own flux. Figure 3 shows that in general response and elasticity coefficients differ. There are, however, a few notable exceptions. The first of these is the trivial case of the first and last metabolites in the chain (A and D): Response coefficients and elasticities will generally agree because the supply block for A and demand block for D each consist only of a single enzyme. (The exception of $\varepsilon_a^{v_1} \neq R_a^{J_1}$ in Figure 3b has to do with the feedback loop and is further discussed in [6]).

Aside from the trivial case, any agreement between elasticity and response coefficient points to a site of regulation. Equation 1 shows that the response coefficient can equal the elasticity either if the control coefficient is one (as discussed above), or if the elasticity is zero (which effectively makes the value of the control coefficient irrelevant). The first case obtains, for example, in Figure 3b with C clamped, where $\varepsilon_c^{v_1} = R_c^{J_{123}}$ (feedback loop with $C_{v_1}^{J_{123}} = 1$). Here, C can be classified as a "regulatory metabolite" with respect to its supply block because the flux response of this supply towards the clamped metabolite concentration is exactly the same as the activity response (i. e. elasticity) of the enzyme directly affected by the clamped metabolite. The flux-control coefficient of one causes the flux response to be transmitted fully through the block.

The second case (zero elasticity) obtains, for example, in Figure 3b, where $\varepsilon_c^{v_4} = R_c^{J_{45}} \approx 0$. Such a zero elasticity confers flux control (in the complete system) to that particular block and results in functional differentiation of the system, which is further discussed below.

FUNCTIONAL DIFFERENTIATION AND HOMEOSTASIS

SDA has shown that when one block (say, demand) controls the flux through a pathway, the other (say, supply) will determine the degree of control of the intermediate [4]. Such a pathway has been termed "functionally differentiated" as flux and concentration control are functions of different blocks. Complete flux control by a supply or demand block (over the whole pathway) can easily be identified by a zero response coefficient (i.e. block elasticity) of that block towards the intermediate (e.g. $R_C^{J_{45}}$ in Fig. 3b). The response coefficient of the other block ($R_C^{J_{123}}$) will then determine the degree of homeostasis in the intermediate: the larger its numerical value, the better the homeostatic buffering.

Model II (Fig. 3b) has been discussed in detail as an example of a functionally differentiated system in the context of SDA [4, 13], and the arguments will not be repeated here. Suffice it to say that the properties of the feedback elasticity $\varepsilon_c^{v_1}$, which equals $R_C^{J_{123}}$ here, set the steady-state concentration of C and determine its degree of homeostatic buffering. In this sense, the steady-state concentration of C can be regarded as "regulated". Why C can be considered a "regulatory" metabolite when considering the supply block in isolation has been discussed above.

Multiple routes of interaction

When two or more direct routes of interaction exist from a clamped metabolite to a particular supply or demand block, GSDA can be further refined by dissecting the response coefficient into partial response coefficients. An example is the GSDA around metabolite C in Figure 1, where C can affect both enzymes 1 and 3 directly (the former through allosteric inhibition, the latter through product inhibition). By calculating the total response coefficient as the sum of the two partial response coefficients referring to each of these routes of interaction, their individual contribution to the total response coefficient can be quantified and depicted graphically. This will not be further discussed here, and the reader is referred to [6].

3 Real Models

The linear pathway in Figure 1 can be regarded as quite artificial, although it illustrates some important regulatory features observed in metabolic pathways. To demonstrate the wider applicability of our approach, we now perform a GSDA on two kinetic models of real pathways from the literature.

3.1 Threonine Biosynthesis in Escherichia coli

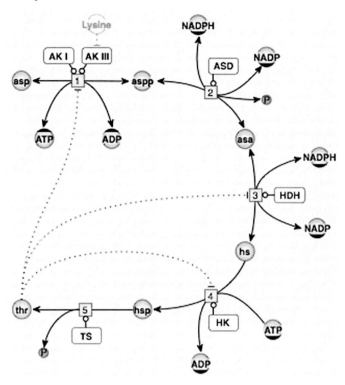

Figure 4. Pathway of threonine biosynthesis in *Escherichia coli* as modelled in [14]. Abbreviations: *asp*, aspartate; *aspp*, β-aspartyl phosphate; *asa*, aspartate β-semialdehyde; *hs*, homoserine; *hsp*, O-phospho-homoserine; *thr*, threonine; *AK*, aspartate kinase; *ASD*, aspartate semialdehyde dehydrogenase; *HDH*, homoserine dehydrogenase; *HK*, homoserine kinase; *TS*, threonine synthase. Image reproduced from JWS Online (http://jjj.biochem.sun.ac.za) with permission.

The first model is of the threonine biosynthesis pathway in *Escherichia coli*, and was developed by Chassagnole *et al.* in 2001 [14] (the pathway structure is given in Fig. 4).

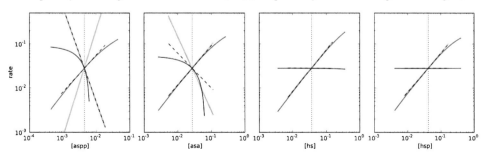

Figure 5. GSDA of the threonine biosynthesis model of [14]. Abbreviations are defined in Fig. 4.

As was done for the model of the linear pathway in Figure 1, each of the intermediates (aspp, asa, hs and hsp) was now in turn made into a parameter of the model and scanned around its reference steady-state value. The results are shown in Figure 5 using the same conventions as previously. Homoserine and phospho-homoserine act as regulatory metabolites in the sense that the flux-response coefficients of their respective supply and demand blocks equal the elasticities of the enzymes directly connected to them. Aspartate-semialdehyde has a similar effect on its demand (homoserine dehydrogenase), but is not regulatory with respect to its supply enzyme, aspartate semialdehyde dehydrogenase-response and elasticity coefficients differ. Aspartyl-phosphate is not a regulatory metabolite (the fact that the elasticity and response coefficients agree for aspartate kinase has to do with the fact that this is the first enzyme in the pathway and the "supply" of aspartyl phosphate only consists of a single enzyme; see Section 2.2 above).

The results can be understood and interpreted as follows. In the model [14] the first two steps (aspartate kinase and aspartate semialdehyde dehydrogenase) are close to equilibrium, leading to large elasticity values for aspartyl phosphate and aspartate semialdehyde. Consequently, elasticities and response coefficients do not agree. Homoserine kinase and threonine synthase (the last two steps) are modelled as irreversible reactions [14], resulting in zero product elasticities. The insensitivity to their products allows these enzymes to transmit any changes in their substrate concentrations downstream, hence the response coefficients and elasticities for these substrates agree (Fig. 5, panels 3 and 4). Homoserine dehydrogenase seems to follow a similar pattern: its substrate elasticity and flux-response coefficient towards aspartate semialdehyde are equal (Fig. 4, panel 2), while its product elasticity is zero (Fig. 4, panel 3). Although the enzyme is modelled with reversible kinetics, the large value of the equilibrium constant (3162, [14]) makes it behave like an effectively irreversible reaction.

3.2 Erythrocyte Glycolysis

The second example concerns a kinetic model of the free-energy and redox metabolism of erythrocytes [15]. For lack of space, we do not show the complete GSDA, but rather focus on the branch-point at glucose-6-phosphate. This metabolite is produced in the hexokinase (HK) reaction, and can then be further metabolised by phosphoglucoisomerase (PGI) in the Embden-Meyerhof-Parnas pathway, or by glucose-6-phosphate dehydrogenase (G6PDH) in the pentose phosphate pathway.

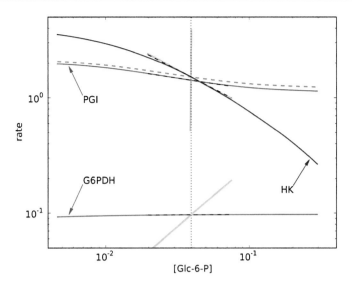

Figure 6. GSDA around the metabolite glucose-6-phosphate in the kinetic model of erythrocyte metabolism of Holzhütter [15]. The graph follows the same convention as previously. For description and discussion see text. Abbreviations: *Glc-6-P*, glucose-6-phosphate; *HK*, hexokinase; *PGI*, phosphoglucoisomerase; *G6PDH*, glucose-6-phosphate dehydrogenase.

The results of the GSDA, performed by clamping the glucose-6-phosphate concentration in the model, varying it around its steady-state value and monitoring the supply (HK) and demand (PGI and G6PDH) fluxes, is shown in Figure 6. This graph differs from Figures 2, 3 and 5 in that it describes a branch-point with two demand fluxes. As discussed in detail in [6], such a case is treated by plotting the sum of the demand fluxes as a green dotted line. The steady state can be read off from the graph where this line intersects the single (blue) supply rate characteristic. The flux-response coefficients and elasticities are plotted for each demand flux separately; in this way the regulatory function of glucose-6-phosphate can be separately assessed for the two branches.

Figure 6 shows that glucose-6-phosphate acts as a regulatory metabolite with respect to its supply (i.e. the elasticity and response coefficient are the same). This is a non-trivial case, since the supply in this model consists of two reactions (glucose transport and HK). However, it can be understood quite readily because glucose transport is modelled as facilitated diffusion and is close to equilibrium in the reference steady state calculated by the model. In contrast to the supply, glucose-6-phosphate is not a regulatory metabolite for either of the demand fluxes (the response and elasticity coefficients differ for both PGI and G6PDH).

Another result from Figure 6, which may appear paradoxical at first, is that the response coefficient of the PGI flux towards glucose-6-phosphate is negative (i.e. increasing glucose-6-phosphate concentrations cause this flux to decrease), although it is a demand flux.

The elasticity of PGI is positive as expected, and moreover has a very large value since the enzyme is close to equilibrium. The reason for the negative response coefficient is that the response of lower glycolysis to glucose-6-phosphate is not only mediated via its effect on PGI, but also through the levels of the ATP/ADP and NAD^+/NADH moieties, which are treated as variables in the model.

4 Concluding Remarks

This paper has described generalised supply-demand analysis as a method for identifying and characterising regulatory metabolites in kinetic models of cellular pathways. The method can be generally applied to complex networks. The approach involves clamping each of the variable species of the model in turn and varying their concentration over a range in a parameter scan. The rate characteristics of supply and demand of that particular species are then generated and plotted, together with straight lines representing the elasticities of the enzymes directly connected to the clamped intermediate, and the response coefficients of the supply and demand blocks. GSDA can pinpoint potential sites of regulation in a pathway, identify regulatory metabolites and sites of functional differentiation, and quantify the importance of different routes of interaction in a pathway [6].

In addition, to our knowledge the results presented in Figures 5 and 6 are the first application of GSDA to kinetic models of "real" pathways and exemplify how the analysis can be used to identify regulatory metabolites from such models.

As discussed in detail in [6], there are obvious inter-relations between GSDA and the "modular" [16] or "top-down" [17, 18] approaches to control analysis, and if there are no additional routes of communication between the supply and demand blocks other than through the intermediate, all enzymes belonging to a particular block (say, supply) form a "monofunctional unit" [19]. GSDA, however, goes further than these "classical" approaches to control analysis: first, by considering the behaviour of the system over a wide range, a broader picture of its control and regulation (e.g. in the face of varying demand loads) is obtained than from a mere set of control and elasticity coefficients at a single steady-state point; and second, the rate characteristics and associated elasticity slopes provide a visual picture that allows easy inference of which block controls the flux, to what extent the intermediate is homeostatically buffered, etc.

In conclusion, the strength of GSDA lies in the fact that it provides a computational tool for the systematic functional analysis of large "silicon-cell"-type kinetic models. The tool has been implemented in the ratechar module of the PySCeS software [20], which has been developed in our group. This provides an easy-to-use general interface that allows the user to perform a GSDA on a model with minimal additional programming. Moreover, since PySCeS can import SBML files [21], models can be imported from public databases such as JWS Online [2] or BioModels [3] and do not need to be re-coded. By including all model species in the analysis, human bias is removed and regulatory metabolites can be readily identified. In subsequent refined analyses, the modeller can then focus on and zoom in on those parts of the model exhibiting interesting regulatory behaviour.

ACKNOWLEDGEMENTS

This work was supported by the National Research Foundation and the National Bioinformatics Network (South Africa).

REFERENCES

[1]   Kitano, H. (2002) Computational systems biology. *Nature* **420**:206 – 210.
       doi: http://dx.doi.org/10.1038/nature01254.

[2]   Olivier, B.G. and Snoep, J.L. (2004) Web-based kinetic modelling using JWS Online.
       Bioinformatics **20**:2143 – 2144.
       doi: http://dx.doi.org/10.1093/bioinformatics/bth200.

[3]   LeNovère, N., Bornstein, B., Broicher, A., Courtot, M., Donizelli, M., Dharuri, H.,
       Li, L., Sauro, H., Schilstra, M., Shapiro, B., Snoep, J.L. and Hucka, M. (2006)
       BioModels Database: a free, centralized database of curated, published, quantitative
       kinetic models of biochemical and cellular systems. *Nucleic Acids Res.* **34**:D 689 –
       D 691.
       doi: http://dx.doi.org/10.1093/nar/gkj092.

[4]   Hofmeyr, J.-H.S. and Cornish-Bowden, A. (2000) Regulating the cellular economy
       of supply and demand. *FEBS Lett.* **476**:47 – 51.
       doi: http://dx.doi.org/10.1016/S0014-5793(00)01668-9.

[5]   Koebmann, B.J., Westerhoff, H.V., Snoep, J.L., Nilsson, D. and Jensen, P.R. (2002)
       The glycolytic flux in *Escherichia coli* is controlled by the demand for ATP. *J.
       Bacteriol.* **184**:3909 – 3916.
       doi: http://dx.doi.org/10.1128/JB.184.14.3909-3916.2002.

[6]   Rohwer, J.M. and Hofmeyr, J.-H.S. (2008) Identifying and characterising regulatory
       metabolites with generalised supply-demand analysis. *J. Theor. Biol.* **252**:546 – 554.
       doi: http://dx.doi.org/10.1016/j.jtbi.2007.10.032.

[7]   Hofmeyr, J.-H.S. (1995) Metabolic regulation: a control analytic perspective.
       J. Bioenerg. Biomembr. **27**:479 – 489.
       doi: http://dx.doi.org/10.1007/BF02110188.

[8]   Kacser, H., Burns, J.A. and Fell, D.A. (1995) The control of flux: 21 years on.
       Biochem. Soc. Trans. **23**:341 – 366.

[9]   Hofmeyr, J.-H.S. and Cornish-Bowden, A. (1997) The reversible Hill equation: how to incorporate cooperative enzymes into metabolic models. *Comp. Appl. Biosci.* **13**:377 – 385

[10]   Fell, D.A. and Sauro, H.M. (1985) Metabolic control and its analysis. Additional relationships between elasticities and control coefficients. *Eur. J. Biochem.* **148**:555 – 561.
       doi: http://dx.doi.org/10.1111/j.1432-1033.1985.tb08876.x.

[11]   Hofmeyr, J.-H.S., Cornish-Bowden, A. and Rohwer, J.M. (1993) Taking enzyme kinetics out of control; putting control into regulation. *Eur. J. Biochem.* **212**:833 – 837.
       doi: http://dx.doi.org/10.1111/j.1432-1033.1993.tb17725.x.

[12]   Hofmeyr, J.-H.S. and Cornish-Bowden, A. (1996) Co-response analysis: A new experimental strategy for metabolic control analysis. *J. Theor. Biol.* **182**:371 – 380.
       doi: http://dx.doi.org/10.1006/jtbi.1996.0176.

[13]   Hofmeyr, J.-H.S. and Cornish-Bowden, A. (1991) Quantitative assessment of regulation in metabolic systems. *Eur. J. Biochem.* **200**:223 – 236.
       doi: http://dx.doi.org/10.1111/j.1432-1033.1991.tb21071.x.

[14]   Chassagnole, C., Fell, D.A., Raÿs, B., Kudla, B. and Mazat, J.-P. (2001) Control of the threonine-synthesis pathway in *Escherichia coli*: a theoretical and experimental approach. *Biochem. J.* **356**:433 – 444.
       doi: http://dx.doi.org/10.1042/0264-6021:3560433.

[15]   Holzhütter, H.-G. (2004) The principle of flux minimization and its application to estimate stationary fluxes in metabolic networks. *Eur. J. Biochem.* **271**:2905 – 2922.
       doi: http://dx.doi.org/10.1111/j.1432-1033.2004.04213.x.

[16]   Schuster, S., Kahn, D. and Westerhoff, H.V. (1993) Modular analysis of the control of complex metabolic pathways. *Biophys. Chem.* **48**:1 – 17.
       doi: http://dx.doi.org/10.1016/0301-4622(93)80037-J.

[17]   Brown, G.C., Hafner, R.P. and Brand, M.D. (1990) A 'top-down' approach to the determination of control coefficients in metabolic control theory. *Eur. J. Biochem.* **188**:321 – 325.
       doi: http://dx.doi.org/10.1111/j.1432-1033.1990.tb15406.x.

[18]   Quant, P.A. (1993) Experimental application of top-down control analysis to metabolic systems. *Trends Biochem. Sci.* **18**:26 – 30.
       doi: http://dx.doi.org/10.1016/0968-0004(93)90084-Z.

[19] Rohwer, J.M., Schuster, S. and Westerhoff, H.V. (1996) How to recognize mono-functional units in a metabolic system. *J. Theor. Biol.* **179**:213 – 228.
doi: http://dx.doi.org/10.1006/jtbi.1996.0062.

[20] Olivier, B.G., Rohwer, J.M. and Hofmeyr, J.-H.S. (2005) Modelling cellular systems with PySCeS. *Bioinformatics* **21**:560 – 561.
doi: http://dx.doi.org/10.1093/bioinformatics/bti046.

[21] Hucka, M., Finney, A., Sauro, H.M., Bolouri, H., Doyle, J.C., Kitano, H., Arkin, A.P., Bornstein, B.J., Bray, D., Cornish-Bowden, A., Cuellar, A.A., Dronov, S., Gilles, E.D., Ginkel, M., Gor, V., Goryanin, I.I., Hedley, W.J., Hodgman, T.C., Hofmeyr, J.-H., Hunter, P.J., Juty, N.S., Kasberger, J.L., Kremling, A., Kummer, U., LeNovère, N., Loew, L.M., Lucio, D., Mendes, P., Minch, E., Mjolsness, E.D., Nakayama, Y., Nelson, M.R., Nielsen, P.F., Sakurada, T., Schaff, J.C., Shapiro, B.E., Shimizu, T.S., Spence, H.D., Stelling, J., Takahashi, K., Tomita, M., Wagner, J. and Wang, J. (2003) The systems biology markup language (SBML): a medium for representation and exchange of biochemical network models. *Bioinformatics* **19**:524 – 531.
doi: http://dx.doi.org/10.1093/bioinformatics/btg015.

THE USE OF *IN VIVO*-LIKE ENZYME KINETICS IN A COMPUTATIONAL MODEL OF YEAST GLYCOLYSIS

KAREN VAN EUNEN[1,2], JOSÉ KIEWIET[1], HANS V. WESTERHOFF[1,2,3] AND BARBARA M. BAKKER[1,2,4,*]

[1]Department of Molecular Cell Physiology, Vrije Universiteit Amsterdam, De Boelelaan 1085, 1081 HV Amsterdam, The Netherlands

[2]Kluyver Centre for Genomics of Industrial Fermentation, Julianalaan 67, 2628 BC Delft, The Netherlands

[3]Manchester Centre for Integrative Systems Biology, Manchester Interdisciplinary BioCentre, The University of Manchester, 131 Princess Street, Manchester M1 7ND, U.K.

[4]Department of Pediatrics, Center for Liver, Digestive and Metabolic Diseases, University Medical Center Groningen, University of Groningen, Hanzeplein 1, 9713 GZ Groningen, The Netherlands

E-Mail: *b.m.bakker@med.umcg.nl

Received: 8th February 2010 / Published: 14th September 2010

ABSTRACT

Usually enzyme kinetic parameters are measured under assay conditions that are optimized for a high activity of the enzyme of interest. The aim of this study was to test if the predictive value of a kinetic computer model of yeast glycolysis would be improved by using kinetic parameters measured in a standardized *in vivo*-like assay medium [1]. The V_{max} and some kinetic parameters of the glycolytic and fermentative enzymes were measured in *Saccharomyces cerevisiae* grown in an aerobic, glucose-limited culture. The assays were done both under '*in-vivo*-like' and optimized conditions. The new data were implemented in an adapted version of the glycolysis model of Teusink *et al.* [2]. The '*in-vivo*-like' enzyme kinetic parameters improved the model substantially as compared to the parameters from optimized assays. In the latter case the model exhibited 'turbo' behaviour, characterized by a dramatic accumulation of hexose phosphates. The *in-vivo*-like kinetic parameters improved the balance between the lower and upper branch of glycolysis and resulted in a better correspondence between model and experiment for both the concentrations of the glycolytic intermediates and the fluxes.

INTRODUCTION

Realistic, quantitative computer models of biochemical networks require that the input data are measured under standardized assay conditions. Moreover, the assay conditions need to be representative of the *in vivo* conditions. However, enzyme-kinetic parameters are most often measured *in vitro* and under optimal conditions for each enzyme. In practice this leads to different assay conditions for each enzyme, *e. g.* in different buffers and at different pH and ionic strength [3 – 5]. In a joint effort of the Dutch Vertical Genomics consortium, the European Yeast Systems Biology Network (YSBN) and the STRENDA (Standards for Reporting Enzymology Data) Commission, we have recently developed an assay medium for measuring enzyme activities that closely resembles the cytosolic environment of the yeast *S. cerevisiae* [1]. Most of the V_{max} values measured in this *in vivo*-like assay medium were lower than those measured under optimal conditions for each enzyme, as one should have expected. The only exceptions were aldolase and pyruvate decarboxylase, which had a higher activity under the *in vivo*-like conditions. The V_{max} values of all enzymes were higher than the flux through them under conditions that favour a high glycolytic flux [1]. This is a prerequisite, since the flux can never exceed the V_{max}. Instead, by sub-saturating substrate concentrations or product inhibition, it can easily drop below the V_{max}. Thus, the new data seem realistic and a good starting point for modelling.

Over the last 30 years many kinetic computer models of yeast glycolysis have been constructed. The first models focused mainly on the mechanisms underlying sustained oscillations in yeast cultures and extracts [6 – 10]. The developments in Metabolic Control Analysis (MCA) inspired the construction of a new generation of models to study the distribution of flux control in glycolysis. The applied aim of these models was to amplify or redirect the flux through glycolysis [11 – 14]. The more recent kinetic models were detailed models based on *in vitro* enzyme kinetics [2, 15, 16] and each of these models was developed with a different aim. The model of Rizzi *et al.* [16] was based on published kinetic mechanisms and affinity constants. The enzyme capacities were fitted on data obtained from dynamic experiments of a glucose pulse added to a steady-state culture. Hynne *et al.* modelled the dynamic characteristics of oscillating yeast cultures to estimate the kinetic parameters [15]. Both approaches aimed at *in vivo* parameter estimation. However, the approach of Hynne *et al.* allowed the estimation of not only the enzyme capacities but also their affinity constants [15].

The objective of the modelling by Teusink *et al.* was to evaluate critically to what extent biochemical knowledge from *in vitro* studies could be used to predict the glycolytic flux and the concentrations of glycolytic intermediates [2]. Their conclusion was that the *in vitro* kinetics could not describe the *in vivo* activity for all of the glycolytic enzymes satisfactorily. A model reduction technique that was extensively applied to the dynamics of yeast glycolysis is the linear-logarithmic (lin-log) kinetics approach [17], which is closely related to mosaic non-equilibrium thermodynamics (MNET) [18]. The lin-log kinetic framework showed that with simplified kinetics and less parameters still good model predictions were obtained [19–21]. This 'minimalist' approach demonstrated the key importance of the feedback and feed forward loops for glycolytic dynamics [22, 23].

The present study builds on the ideas developed by Teusink *et al.* [2], who used computer modelling as a stringent test of biochemistry. Here we have measured the kinetics of the glycolytic and fermentative enzymes in yeast under *in vivo*-like assay conditions as described in Van Eunen *et al.* [1]. The obtained V_{max} values and some new affinity constants were inserted in the glycolysis model of Teusink *et al.* [2] and it was tested to what extent the *in vivo*-like enzyme kinetics improved the modelling results.

MATERIAL AND METHODS

Growth and sampling

The haploid, prototrophic *Saccharomyces cerevisiae* strain CEN.PK113–7D (*MATa*, *MAL2–8ᶜ*, *SUC2*, obtained from P. Kötter, Frankfurt, Germany) was cultivated in an aerobic glucose-limited chemostat culture at 30 °C as described in [24]. The dilution rate and hence the specific growth rate of the culture was set at 0.35 h⁻¹. When the culture was at steady state, samples were taken to measure the V_{max} values of all the glycolytic and fermentative enzymes. Subsequently the cells were transferred to anaerobic conditions at a high glucose concentration to measure the maximal glycolytic flux and the intracellular metabolite concentrations. The V_{max} values measured under optimized assay conditions, the flux and metabolite data have been reported before [32].

Glucose-transport activity assay

Zero-trans influx of ¹⁴C-labeled glucose was measured in a 5-s uptake assay described by Walsh *et al.* [25] with the modifications of Rossell *et al.* [26] at 30 °C. The range of glucose concentrations was between 0.25 and 225 mM. Irreversible Michaelis-Menten equations were fitted to the data by nonlinear regression.

V_{max} measurements under in vivo-*like assay conditions*

Cell-free extracts were prepared freshly by the FastPrep® method described by Van Eunen *et al.* [1]. V_{max} assays were carried out with the prepared extracts via NAD(P)H-linked assays, at 30 °C in a Novostar spectrophotometer (BMG Labtech) as described in detail in [1].

The standardized *in vivo*-like assay medium [1] contained 300 mM potassium, 245 mM glutamate, 50 mM phosphate, 20 mM sodium, 2 mM free magnesium, 5 – 10 mM sulphate, and 0.5 mM calcium. For the addition of magnesium, it was taken into account that ATP, ADP, NADP and TPP bind magnesium with a high affinity. The amount of magnesium added equalled the summed concentration of these coenzymes plus 2 mM, such that the free magnesium concentration was 2 mM. Since the sulfate salt of magnesium was used, the sulfate concentration in the final assay medium varied in a range between 2.5 and 10 mM. Concentrations of substrates and coupling enzymes were as follows:

Hexokinase (HXK, EC 2.7.1.1) – 1 mM NADP, 10 mM glucose, 1 mM ATP and 1.8 U/ml glucose-6-phosphate dehydrogenase (G6PDH, EC 1.1.1.49).

Phosphoglucose isomerase (PGI, EC 5.3.1.9, reverse direction) – 0.4 mM NADP, 2 mM Fructose 6-phosphate (F6P) and 1.8 U G6PDH.

Phosphofructokinase (PFK, EC 2.7.1.11) – 0.1 mM fructose 2,6-bisphosphate, 0.15 mM NADH, 0.5 mM ATP, 10 mM F6P, 0.45 U/ml aldolase (ALD, EC 4.1.2.13), 0.6 U/ml Glycerol 3-phosphate dehydrogenase (G3PDH, EC 1.1.1.8) and 1.8 U/ml triosephosphate isomerase (TPI, EC 5.3.1.1).

Aldolase – 0.15 mM NADH, 2 mM fructose 1,6-bisphosphate (F16BP), 0.6 U/ml G3PDH and 1.8 U/ml TPI.

Glyceraldehyde 3-phosphate dehydrogenase (GAPDH, EC 1.2.1.12, reverse direction) – 1 mM ATP, 0.15 mM NADH, 5 mM 3-phosphoglyceric acid (3PG) and 22.5 U/ml 3-phos-phoglycerate kinase (PGK, EC 2.7.2.3).

Glyceraldehyde 3-phosphate dehydrogenase (GAPDH, forward direction) – 10 mM ADP, 1 mM NAD, 5.8 mM glyceraldehyde 3-phosphate and 22.5 U/ml PGK.

3-Phosphoglycerate kinase (reverse direction) – 1 mM ATP, 0.15 mM NADH, 5 mM 3PG and 8 U/ml GAPDH.

Phosphoglycerate-mutase (GPM, EC 5.4.2.1) – 10 mM ADP, 0.15 mM NADH, 1.25 mM 2,3-diphospho-D-glyceric acid, 5 mM 3PG, 2 U/ml enolase (ENO, EC 4.2.1.11), 13 U/ml pyruvate kinase (PYK, EC 2.7.1.40) and 11.3 U/ml lactate dehydrogenase (LDH, EC 1.1.1.27).

Enolase – The activity of ENO was measured by the production rate of PEP, which was analyzed with a spectrophotometer using a wavelength of 240 nm. The assay was measured using the *in vivo*-like assay medium with 6 mM of 2-phosphoglyceric acid (2PG).

Pyruvate kinase – 10 mM ADP, 0.15 mM NADH, 1 mM F16BP, 2 mM phosphoenolpyruvate (PEP) and 13.8 U/ml LDH.

Pyruvate decarboxylase (PDC, EC 4.1.1.1) – 0.2 mM TPP, 0.15 mM NADH, 50 mM pyruvate and 88 U/ml alcohol dehydrogenase (ADH, EC 1.1.1.1).

Alcohol dehydrogenase – 1 mM NAD and 100 mM ethanol.

The affinity constants (K_m) of GAPDH for glyceraldehyde 3-phosphate, NAD and NADH were redetermined in the *in vivo*-like assay medium by varying the substrate concentrations.

Model description

The glycolytic model of Teusink *et al.* [2] was the starting point for this study. The aim of the modelling was *(i)* to predict the steady-state flux and metabolite concentrations, under the conditions of the fermentative capacity assay (high glucose, anaerobic) and at the measured V_{max} values, *(ii)* to compare the model outcome to the measured flux and metabolite concentrations and *(iii)* to test how the *in vivo*-like V_{max} values affected the correspondence between model and experiment, as compared to the V_{max} values from the optimized assays. The model as it was used here, is depicted in Figure 1. Starting from the original Teusink model [2] the following modifications were made, based on new insights and in order to tailor the model to the experimental conditions of this study.

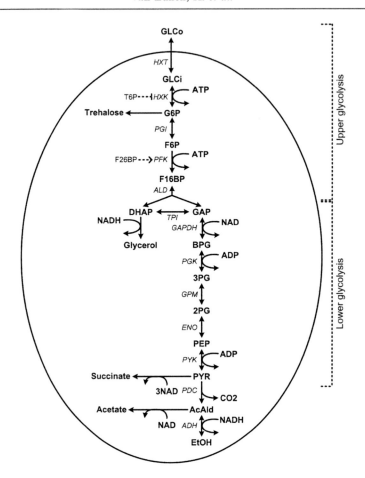

Figure 1. The glycolytic and fermentative pathway as they were modelled in this study. Metabolites are depicted in bold, allosteric regulators in regular, enzymes in italics and branching pathways underlined. GLCo: extracellular glucose, GLCi: intracellular glucose, G6P: glucose 6-phosphate, F6P: fructose 6-phosphate, F16BP: fructose 1,6-bisphosphate, DHAP: dihydroxyacetone phosphate, GAP: glyceraldehyde phosphate, BPG: 1,3-bisphosphoglycerate, 3PG: 3-phosphoglycerate, 2PG: 2-phosphoglycerate, PEP: phosphoenolpyruvate, PYR: pyruvate, ACE: acetaldehyde, EtOH: ethanol, HXT: hexose transport, HXK: hexokinase, PGI: phosphoglucose isomerase, PFK: phosphofructo kinase, ALD: aldolase, TPI: triose-phosphate isomerase, GAPDH: glyceraldehyde 3-phosphate dehydrogenase, PGK: 3-phosphoglycerate kinase, GPM: phosphoglycerate mutase, ENO: enolase, PYK: pyruvate kinase, PDC: pyruvate decarboxylase, ADH: alcohol dehydrogenase.

1. The V_{max} values of all glycolytic and fermentative enzymes and the V_{max} and affinity constant of glucose transport that were measured under the conditions of this study (Table 2) were implemented in the model. For most of the remaining kinetic parameters we have used the values of Teusink *et al.* [2]. The only

exceptions were the kinetic parameters of GAPDH that were re-determined under the *in vivo* like conditions (Table 3). The original GAPDH parameters were used together with the optimized V_{max}, while the newly determined parameters were used together with the *in vivo*-like V_{max}.

2. In the original Teusink model the branching fluxes to trehalose and glycogen were fixed at their measured values. Under the conditions described here, the glycogen flux was negligible and therefore not included. The trehalose flux was fixed at the value measured in this study and so were the fluxes to glycerol and succinate (Table 1). To prevent a redox problem in the model we did not fix the flux to acetate. Instead it was made linearly dependent on the acetaldehyde concentration with a rate constant of 0.5.

3. In the original model the net ATP produced by glycolysis was consumed in a lumped reaction of ATP utilization. This resulted in variable and interdependent ATP, AMP and ADP concentrations. Since information about the kinetics of ATP utilization was lacking and moreover not the focus of this study, we decided to remove the ATP utilization from the model and instead inserted the concentrations of the ATP, ADP and AMP as fixed parameters.

4. The known inhibition of HXK by trehalose-6-phosphate (T6P) was not included in the original model. Yet, it is thought to play an important role in the regulation of glycolysis, particularly to prevent an unbalance between the upper and lower part of the pathway [27]. T6P is an inhibitor of HXK that competes with its substrate glucose. Different K_i values for the different hexokinases of yeast have been reported. Glucokinase was not inhibited by T6P, while the K_i values for hexokinase I and II were 0.2 mM and 0.04 mM, respectively [28]. Here we have used a K_i value of 0.2 mM. The modified kinetic equation of HXK is:

$$v_{hxk} = \cfrac{V_{max,hxk} \cdot \left(\cfrac{Glc_i(t)}{K_{m,hxk,Glci}} \cdot \cfrac{ATP}{K_{m,hxk,ATP}} - \cfrac{G6P(t) \cdot ADP}{K_{m,hxk,Glci} \cdot K_{m,hxk,ATP} \cdot K_{eq,hxk}} \right)}{\left(1 + \cfrac{Glc_i(t)}{K_{m,hxk,Glci}} + \cfrac{G6P(t)}{K_{m,hxk,G6P}} + \cfrac{T6P}{K_{i,hxk,T6P}} \right) \cdot \left(1 + \cfrac{ATP}{K_{m,hxk,ATP}} + \cfrac{ADP}{K_{m,hxk,ADP}} \right)}$$

5. Finally, the K_m of PDC in the original model [2] was obtained from Boiteux and Hess [29] based on a intracellular phosphate concentration of 25 mM. However, we have measured the PDC activity at a phosphate concentration of 50 mM, which is likely to be the intracellular concentration under the growth conditions studied here [30]. Based on the data of Boiteux and Hess [29] we calculated a K_m value of PDC for pyruvate of 6.36 mM at 50 mM phosphate, and the new value was inserted in the model. For comparison, in the original model the K_m of PDC was 4.3 mM.

All experimental data were converted to intracellular units (mM min^{-1} for rates and mM for concentrations) by assuming a yeast cytosolic volume of 3.75 µl.mg cell protein^{-1} [31].

RESULTS

Measurements of glycolytic flux, intracellular metabolite concentrations and V_{max}

Yeast cells were grown under aerobic glucose-limited chemostat conditions at a growth rate of $0.35\,h^{-1}$. When the chemostat cultures were at steady state, cells were harvested to measure *(i)* the maximal glycolytic flux and the intracellular metabolite concentrations in an off-line assay under anaerobic glucose-excess conditions (fermentative capacity) and *(ii)* the V_{max} of the glycolytic and fermentative enzymes. The fluxes and metabolite concentrations, as well as the V_{max} values under optimized assay conditions, were already reported [32] and are taken from the latter study. The V_{max} values under *in vivo*-like conditions and the kinetics of glucose transport are new measurements made for the present study. Tables 1 to 4 show the complete dataset.

Table 1. The measured fluxes through to the side branches, *i. e.* to trehalose, to glycerol and to succinate. The fluxes to these side branches are used as fixed parameters in the model. The data were taken from [32].

	Flux (mM.min^{-1})
Trehalose	1.0 ± 0.3
Glycerol	21.3 ± 0.7
Succinate	0.9 ± 0.0

Table 2. Kinetic parameters measured under optimal and *in vivo*-like assay conditions. Parameters measured under the optimal assay conditions were taken from [32]. The glucose-transport kinetic parameters were measured in intact cells. Since the cells were incubated under the same conditions as used for flux measurements (see Materials and Methods), the results are listed under '*in vivo*-like'. However, the same data for glucose transport were used in both model versions.

Parameter	Optimal assay conditions	*In vivo*-like assay conditions	
$V_{max,glt}$		201.3	mM.min^{-1}
$K_{m,glt,GLC}$		0.9	mM
$V_{max,hxk}$	551.9	257.5	mM.min^{-1}
$V_{max,pgi}$	1141.3	903.4	mM.min^{-1}
$V_{max,pfk}$	98.4	178.7	mM.min^{-1}
$V_{max,ald}$	251.0	200.0	mM.min^{-1}
$V^{+}_{max,gapdh}$	197.3[a]	156[a]/1496[b]	mM.min^{-1}
$V_{max,gapdh}$	1101.0	866.7	mM.min^{-1}
$V_{max,pgk}$	1662.0	2415.5	mM.min^{-1}
$V_{max,gpm}$	1502.7	870.7	mM.min^{-1}
$V_{max,eno}$	285.3	485.2	mM.min^{-1}
$V_{max,pyk}$	965.4	677.3	mM.min^{-1}
$V_{max,pdc}$	218.8	334.7	mM.min^{-1}
$V_{max,adh}$	437.6	856.0	mM.min^{-1}

[a] Forward V_{max} values of GAPDH were not measured but calculated with the Haldane relationship using the kinetic parameters from [2].
[b] Forward V_{max} value was measured.

Table 2 shows the V_{max} of the glycolytic and fermentative enzymes measured under both the optimal and the *in vivo*-like assay conditions. In the case of GAPDH, the *in vivo*-like V_{max} was measured in both directions and also the affinity constants were redetermined under *in vivo*-like condition. Since 1,3-bisphoshoglycerate (BPG) is not stable, we could not measure the affinity constant (K_m) for this product. However, we could calculate the K_m for BPG using the Haldane relationship, the measured V_{max}, the measured K_m values for the other substrates and products, and the K_{eq} (Table 3).

Table 3. Kinetic parameters of GAPDH according to Teusink *et al.* [2] and newly measured under the *in vivo*-like assay conditions.

Parameter	According to Teusink *et al.* [2]	Redetermined under *in vivo*-like assay conditions	
$V^+_{max,gapdh}$	197.3[a]	1496[b]	mM.min^{-1}
$V_{max,gapdh}$	1101.0	866.7	mM.min^{-1}
$K_{m,GAP}$	0.21	0.39	mM
$K_{m,NAD}$	0.09	2.85	mM
$K_{m,NADH}$	0.0098	0.007	mM
$K_{m,BPG}$	0.06	0.51	mM
$K_{eq,gapdh}$	0.0056	0.0056	Taken from [2]

[a] Forward V_{max} values of GAPDH were not measured but calculated with the Haldane relationship using the kinetic parameters from [2].
[b] Forward V_{max} value was measured.

Table 4. Measured intracellular concentrations of the adenine nucleotides and the allosteric regulators of glycolysis that were included in the model. The data were taken from [32].

Metabolite	Intracellular concentration (mM)
ATP	4.29 ± 0.13
ADP	1.29 ± 0.05
AMP	0.44 ± 0.06
T6P	3.52 ± 0.36
F26P	0.004 ± 0.003

The samples for intracellular metabolite measurements had been taken from the incubations for the off-line flux measurements at 15 minutes after the addition of glucose. At that time the measured fluxes had become constant and glycolysis was assumed to be at steady state. The concentrations of the allosteric regulators and those of ATP, ADP and AMP (Table 4) were inserted as fixed parameters in the kinetic model, while the concentrations of the glycolytic intermediates were used as a validation of the model (Table 5).

Table 5. Model results compared to the experimental data. The experimental data were taken from [32].

	Experiment	Model	
		Optimal	*In vivo* New GAPDH parameters
Flux			
HXT-HXK	120 ± 6	84	90
PGI-ALD	99 ± 6	82	88
GAPDH-ADH	177 ± 11	143	155
Metabolites			
G6P	5.4 ± 0.2	**1436.9**	1.9
F6P	1.0 ± 0.0	**378.6**	0.4
F16bP	27 ± 2	**305.6**	27
P3G+P2G	1.2 ± 0.1	0.9	1.2
PEP	0.11 ± 0.01	0.13	0.27
PYR	5.3 ± 0.0	8.7	5.8

Modelling of glycolysis

In order to test the impact of the *in vivo*-like V_{max} values, the V_{max} data measured under both assay conditions were implemented respectively in the adapted kinetic model of yeast glycolysis. Other modifications to the original Teusink model are listed in Materials and Methods.

Figure 2A, C, E and G and the second column of Table 5 show the simulation results of the glycolysis model with the V_{max} data measured under the assay conditions optimized for each enzyme. The model yielded a steady state, however, only at extremely high concentrations for the hexose phosphates in the upper part of glycolysis (Table 5). The concentrations of the metabolites in the lower part of glycolysis came much closer to the experimental data (Table 5).

When the *in vivo*-like V_{max} data were inserted into the model, but not the newly measured *in vivo*-like GAPDH parameters, no steady state was found (data not shown). Fructose 1,6-bisphosphate accumulated, indicating that the lower part of glycolysis failed to keep up with the flux through the upper part of glycolysis. We suspected that this might be due to the GAPDH kinetics. GAPDH is the first enzyme of the lower part of glycolysis and further-more the kinetics of GAPDH already gave problems in the original model of Teusink *et al.* [2]. Therefore the GAPDH parameters that were re-determined under *in vivo*-like conditions (Table 3) were inserted in the model.

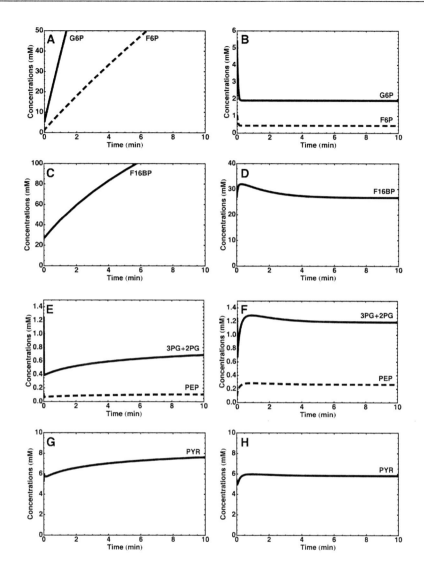

Figure 2. The comparison of the model results of yeast glycolysis obtained with the enzyme-kinetic data measured either in an assay medium optimized for each enzyme (panel A, C, E, and G) or in an assay medium resembling the in vivo situation (panel B, D, F, and H). The concentrations at time zero equal the measured intracellular concentrations. G6P: glucose-6-phosphate, F6P: fructose-6-phosphate, F16BP: fructose-1,6-bisphosphate, 3PG: 3-phosphoglycerate, 2PG: 2-phosphoglycerate, PEP: phosphoenolpyruvate, PYR: pyruvate.

Figure 2B, D, F and H and the last column of Table 5 show the simulation results of the glycolysis model with the *in vivo* V_{max} data and including the new parameters for GAPDH. The steady-state data calculated by the model showed much more similarity with the experimental data. The fluxes deviated by some 20% between model and experiment (Table 5), yet the best correspondence was reached with the *in vivo*-like V_{max} values. As compared to the model version with optimized kinetic parameters, the most substantial improvement was made in the concentrations of the hexose phosphates in the upper part of glycolysis. They came back to the experimentally observed range and the balance between upper and lower glycolysis was restored.

CONCLUSION

This paper showed the importance of a standardized enzyme-assay medium which represents the *in vivo* conditions, for realistic quantitative models. In a joint effort of the Dutch Vertical Genomics consortium, the European Yeast Systems Biology Network (YSBN) and the STRENDA (Standards for Reporting Enzymology Data) Commission, such an assay medium was developed for the determination of enzyme-kinetic parameters in the yeast *S. cerevisiae* [1].

The implementation of the *in vivo*-like V_{max} values and the re-determined kinetic parameters of GAPDH has led to an improvement of the glycolytic model. Both fluxes and intracellular concentrations of the glycolytic intermediates showed similarity with the experimental data. The redetermination of the affinity constants of GAPDH was essential, demonstrating that not only V_{max} values but also other kinetic parameters need to be evaluated under the *in vivo*-like assay conditions in future research.

We also tested the impact of the *in vivo*-like parameters when the yeast was grown and maintained under other conditions. Then the correspondence between model and experiment was not quite as good as under the conditions of the present paper (not shown). In ongoing studies it is tested whether redetermination of a larger set of kinetic parameters under the *in vivo*-like conditions will improve the model.

ACKNOWLEDGEMENTS

This project was supported financially by the IOP Genomics program of Senter Novem. The work of B.M. Bakker and H.V. Westerhoff is further supported by a Rosalind Franklin Fellowship to B.M. Bakker, STW, NGI-Kluyver Centre, NWO-SysMO, BBSRC (including SysMO), EPSRC, AstraZeneca, and EU grants BioSim, NucSys, ECMOAN, and UniCellSys. The CEN.PK113 – 7D strain was kindly donated by P. Kötter, Euroscarf, Frankfurt.

REFERENCES

[1]   van Eunen, K., Bouwman, J., Daran-Lapujade, P., Postmus, J., Canelas, A.B., Mensonides, F.I., Orij, R., Tuzun, I., van den Brink, J., Smits, G.J., van Gulik, W.M., Brul, S., Heijnen, J.J., de Winde, J.H., Teixeira de Mattos, M.J., Kettner, C., Nielsen, J., Westerhoff, H.V. and Bakker, B.M. (2010) Measuring enzyme activities under standardized *in vivo*-like conditions for systems biology. *FEBS J.* **277**(3):749 – 760. doi: http://dx.doi.org/10.1111/j.1742-4658.2009.07524.x.

[2]   Teusink, B., Passarge, J., Reijenga, C.A., Esgalhado, E., van der Weijden, C.C., Schepper, M., Walsh, M.C., Bakker, B.M., van Dam, K., Westerhoff, H.V. and Snoep, J.L. (2000) Can yeast glycolysis be understood in terms of *in vitro* kinetics of the constituent enzymes? Testing biochemistry. *Eur. J. Biochem.* **267**(17):5313 – 5329. doi: http://dx.doi.org/10.1046/j.1432-1327.2000.01527.x.

[3]   Even, S., Lindley, N.D. and Cocaign-Bousquet, M. (2001) Molecular physiology of sugar catabolism in *Lactococcus lactis* IL 1403. *J. Bacteriol.* **183**(13):3817 – 3824. doi: http://dx.doi.org/10.1128/JB.183.13.3817-3824.2001.

[4]   Postma, E., Verduyn, C., Scheffers, W.A. and Van Dijken, J.P. (1989) Enzymic analysis of the crabtree effect in glucose-limited chemostat cultures of *Saccharomyces cerevisiae*. *Appl. Environ. Microbiol.* **55**(2):468 – 477.

[5]   Van Hoek, P., Van Dijken, J.P. and Pronk, J.T. (1998) Effect of specific growth rate on fermentative capacity of baker's yeast. *Appl. Environ. Microbiol.* **64**(11):4226 – 4233.

[6]   Betz, A. and Chance, B. (1965) Phase relationship of glycolytic intermediates in yeast cells with oscillatory metabolic control. *Arch. Biochem. Biophys.* **109**:585 – 594. doi: http://dx.doi.org/10.1016/0003-9861(65)90404-2.

[7]   Boiteux, A., Goldbeter, A. and Hess, B. (1975) Control of oscillating glycolysis of yeast by stochastic, periodic, and steady source of substrate: a model and experimental study. *Proc. Natl. Acad. Sci. U.S.A.* **72**(10):3829 – 3833. doi: http://dx.doi.org/10.1073/pnas.72.10.3829.

[8]   Boiteux, A. and Busse, H.G. (1989) Circuit analysis of the oscillatory state in glycolysis. *Biosystems* **22**(3):231 – 240. doi: http://dx.doi.org/10.1016/0303-2647(89)90064-6.

[9]   Hess, B. and Boiteux, A. (1968) Mechanism of glycolytic oscillation in yeast. I. Aerobic and anaerobic growth conditions for obtaining glycolytic oscillation. *Hoppe Seylers Z. Physiol. Chem.* **349**(11):1567 – 1574.

[10]   Richter, O., Betz, A. and Giersch, C. (1975) The response of oscillating glycolysis to perturbations in the NADH/NAD system: a comparison between experiments and a computer model. *Biosystems* **7**(1):137–146.
doi: http://dx.doi.org/10.1016/0303-2647(75)90051-9.

[11]   Cortassa, S. and Aon, M.A. (1994) Metabolic control analysis of glycolysis and branching to ethanol production in chemostat cultures of *Saccharomyces cerevisiae* under carbon, nitrogen, or phosphate limitations. *Enzyme and Microbial Technology* **16**(9):761–770.
doi: http://dx.doi.org/10.1016/0141-0229(94)90033-7.

[12]   Delgado, J., Meruane, J. and Liao, J.C. (1993) Experimental determination of flux control distribution in biochemical systems: *In vitro* model to analyze transient metabolite concentrations. *Biotechnol. Bioeng.* **41**(11):1121–1128.
doi: http://dx.doi.org/10.1002/bit.260411116.

[13]   Schlosser, P.M., Riedy, T.G. and Bailey, J.E. (1994) Ethanol-production in bakers-yeast – Application of experimental perturbation techniques for model development and resultant changes in Flux Control Analysis. *Biotechnology Progress* **10**(2):141–154.
doi: http://dx.doi.org/10.1021/bp00026a003.

[14]   Galazzo, J.L. and Bailey, J.E. (1990) Fermentation pathway kinetics and metabolic flux control in suspended and immobilized *Saccharomyces cerevisiae*. *Enzyme and Microbial Technology* **12**(3):162–172.
doi: http://dx.doi.org/10.1016/0141-0229(90)90033-M.

[15]   Hynne, F., Dano, S. and Sorensen, P.G. (2001) Full-scale model of glycolysis in *Saccharomyces cerevisiae*. *Biophys. Chem.* **94**(1–2):121–163.
doi: http://dx.doi.org/10.1016/S0301-4622(01)00229-0.

[16]   Rizzi, M., Baltes, M., Theobald, U. and Reuss, M. (1997) In vivo analysis of metabolic dynamics in *Saccharomyces cerevisiae*: II. Mathematical model. *Biotechnol. Bioeng.* **55**(4):592–608.
doi: http://dx.doi.org/10.1002/(SICI)1097-0290(19970820)55:4<592::AID-BIT2>3.3.CO;2-1.

[17]   Visser, D. and Heijnen, J.J. (2003) Dynamic simulation and metabolic re-design of a branched pathway using linlog kinetics. *Metab. Eng.* **5**(3):164–176.
doi: http://dx.doi.org/10.1016/S1096-7176(03)00025-9.

[18]   Westerhoff, H.V. and van Dam, K. (1987) *Thermodynamics and control of biological free energy transduction.* Elsevier, Amsterdam.

[19] Kresnowati, M.T., van Winden, W.A. and Heijnen, J.J. (2005) Determination of elasticities, concentration and flux control coefficients from transient metabolite data using linlog kinetics. *Metab. Eng.* **7**(2):142–153.
doi: http://dx.doi.org/10.1016/j.ymben.2004.12.002.

[20] Visser, D., Schmid, J.W., Mauch, K., Reuss, M. and Heijnen, J.J. (2004) Optimal re-design of primary metabolism in *Escherichia coli* using linlog kinetics. *Metab. Eng.* **6**(4):378–390.
doi: http://dx.doi.org/10.1016/j.ymben.2004.07.001.

[21] Wu, L., Mashego, M.R., Proell, A.M., Vinke, J.L., Ras, C., van Dam, J., van Winden, W.A., van Gulik, W.M. and Heijnen, J.J. (2006) *In vivo* kinetics of primary metabolism in *Saccharomyces cerevisiae* studied through prolonged chemostat culti-vation. *Metab. Eng.* **8**(2):160–171.
doi: http://dx.doi.org/10.1016/j.ymben.2005.09.005.

[22] Nikerel, I.E., van Winden, W.A., van Gulik, W.M. and Heijnen, J.J. (2006) A method for estimation of elasticities in metabolic networks using steady state and dynamic metabolomics data and linlog kinetics. *BMC Bioinformatics* 7540.

[23] Rossell, S., Solem, C., Verheijen, P.J.T., Jensen, P.R. and Heijnen, J.J. (2008) Approximate flux functions. In *International Study Group for Systems Biology: Will bottom up meet top down?* (Edited by Hansen, A.C.H., Koebmann, B. & Jensen, P.R.), pp. 61–72, Elsinore, Denmark.

[24] van Eunen, K., Bouwman, J., Lindenbergh, A., Westerhoff, H.V. and Bakker, B.M. (2009) Time-dependent regulation analysis dissects shifts between metabolic and gene-expression regulation during nitrogen starvation in baker's yeast. *FEBS J.* **276**(19):5521–5536.
doi: http://dx.doi.org/10.1111/j.1742-4658.2009.07235.x.

[25] Walsh, M.C., Smits, H.P., Scholte, M. and van Dam, K. (1994) Affinity of glucose transport in *Saccharomyces cerevisiae* is modulated during growth on glucose. *J. Bacteriol.* **176**(4):953–958.

[26] Rossell, S., van der Weijden, C.C., Kruckeberg, A.L., Bakker, B.M. and Wester-hoff, H.V. (2005) Hierarchical and metabolic regulation of glucose influx in starved *Saccharomyces cerevisiae*. *FEMS Yeast Res.* **5**(6–7):611–619.
doi: http://dx.doi.org/10.1016/j.femsyr.2004.11.003.

[27] Teusink, B., Walsh, M.C., van Dam, K. and Westerhoff, H.V. (1998) The danger of metabolic pathways with turbo design. *Trends Biochem. Sci.* **23**(5):162–169.
doi: http://dx.doi.org/10.1016/S0968-0004(98)01205-5.

[28]   Blazquez, M.A., Lagunas, R., Gancedo, C. and Gancedo, J.M. (1993) Trehalose-6-phosphate, a new regulator of yeast glycolysis that inhibits hexokinases. *FEBS Lett.* **329**(1–2):51–54.
doi: http://dx.doi.org/10.1016/0014-5793(93)80191-V.

[29]   Boiteux, A. and Hess, B. (1970) Allosteric properties of yeast pyruvate decarboxylase. *FEBS Lett.* **9**(5):293–296.
doi: http://dx.doi.org/10.1016/0014-5793(70)80381-7.

[30]   Wu, L., van Dam, J., Schipper, D., Kresnowati, M.T., Proell, A.M., Ras, C., van Winden, W.A., van Gulik, W.M. and Heijnen, J.J. (2006) Short-term metabolome dynamics and carbon, electron, and ATP balances in chemostat-grown *Saccharomyces cerevisiae* CEN.PK 113–7D following a glucose pulse. *Appl. Environ. Microbiol.* **72**(5):3566–3577.
doi: http://dx.doi.org/10.1128/AEM.72.5.3566-3577.2006.

[31]   de Koning, W. and van Dam, K. (1992) A method for the determination of changes of glycolytic metabolites in yeast on a subsecond time scale using extraction at neutral pH. *Anal. Biochem.* **204**(1):118–123.
doi: http://dx.doi.org/10.1016/0003-2697(92)90149-2.

[32]   van Eunen, K., Dool, P., Canelas, A.B., Kiewiet, J., Bouwman, J., van Gulik, W.M., Westerhoff., H.V. and Bakker, B.M. (2010) Time-dependent regulation of yeast glycolysis upon nitrogen starvation depends on cell history. *IET Syst. Biol.* **4**(2):157–168.
doi: http://dx.doi.org/10.1049/iet-syb.2009.0025.

Developing Coherent Minimum Reporting Guidelines for Biological Scientists: The MIBBI Project

Chris F. Taylor

European Bioinformatics Institute, Wellcome Trust Genome Campus, Hinxton, Cambridge, CB10 1SD, U.K.

E-Mail: chris.taylor@ebo.ac.uk

Received: 17th March 2010 / Published: 14th September 2010

Abstract

Modern biological science addresses a variety of subjects using an array of analytical techniques. Few relations between subject and technique are exclusive, making for a very large number of potential workflows, combinatorially-speaking. While this diversity is to be celebrated, it presents informatics challenges that require resolution if the data-sharing ambitions of many funders are to be realised, and the consequent benefits to science obtained. There is increasingly-organised movement towards consensus on data and reporting standards for the biosciences, but significant hurdles remain: scientists must be convinced of the value of the exercise, and user-friendly, time-efficient and robust tools are required. The 'Minimum Information for Biological and Biomedical Investigations' (MIBBI) Project, which promotes and develops guidance on the content that experimental reports should contain, is dependent on progress in both these areas.

Introduction

In recent years the suffix 'omics' has come rather too much into use. While conveying some information about the intent of an investigation (for example, 'proteomics' indicating the study of the protein complement of a [biological] sample), the various omics-terminated terms came more commonly to connote sets of associated technologies (for proteomics, mass spectrometry). For reasons both sociological and technological, nascent societies associated with these emerging omics fields spawned standards bodies tasked to smooth out the wrinkles in their data sharing landscape, but those standards bodies' scopes overlapped (for example, mass spectrometry has many uses beyond proteomics), resulting in redundancy of effort. Coordination became essential; not just amongst omics fields, but with the wider bioscience community, within which similar efforts were progressing for a variety of areas, focused on particular fields/workflows.

Given the innumerable combinations of areas of enquiry and techniques of investigation within the biosciences, an appropriate approach is to develop modular solutions (*i.e.*, separable parts; whether of a vocabulary, format or set of guidelines). The Ontology of Biomedical Investigations (OBI; http://purl.obolibrary.org/obo/obi), and at a higher level the Open Biomedical Ontologies project ([1]; http://obofoundry.org/) of which it is part, reflect this approach; as do the structures of data exchange formats such as FuGE [2] and ISA-Tab [3]. Such modules require a general framework to define and contextualise them; if that framework is shared between resources their combined use is simplified. The framework used by an increasing number of projects, including ISA-Tab and the MIBBI Foundry (described later) consists most importantly of three concepts: *Investigation*, *Study* and *Assay* (Figure 1). In this simple framework, an *Investigation* is defined as a body of work consisting of one or more *Studies* (which normally centre on a biological question or a particular source material), each of which may contain one or more *Assays* (usually some kind of data-generating analysis).

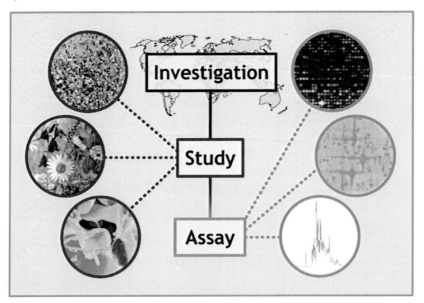

Figure 1. A graphical representation of the Investigation/Study/Assay (ISA) hierarchy, used as a basic framework for the data structures underlying an increasing number of standardisation projects such as MIBBI and ISA-Tab.

ON COORDINATION

There are various efforts to coordinate the development of standards (OBI/OBO, MIBBI, FuGE, ISA-Tab, *etc.*; [4]). The situation is still in flux, but promises to continue to settle down as patterns and precedents become more established. In recent years, one significant driver has been policy development by funding agencies. Their various statements and regulations relating to data sharing [5] all parallel statements made by the Organisation for Economic Cooperation and Development (OECD; http://oecd.org/), which holds that publicly-funded research data are a public good, produced in the public interest, and should be openly available to the maximum extent possible.

Broadly, there are three kinds of standard: data formats such as ISA-Tab; vocabularies/ontologies such as those in OBO; and Minimum Information (MI) checklists, which are the focus of this paper. MI checklists are guidance documents specifying the information that should be provided when reporting research work. Historically, these have been developed by groups working within particular biological or technological domains. The resulting fragmentation made it difficult to obtain an overview of ongoing projects, or to track progress. Furthermore, having been developed independently, checklists were frequently partially redundant against each other, and arbitrary decisions on wording and substructuring in the overlaps ensured that integration would be laborious. This made it impractical for potential users to combine parts of checklists to make guidance documents appropriate for their own workflow.

THE MIBBI PROJECT

The Minimum Information for Biological and Biomedical Investigations (MIBBI) Project is an international collaboration between communities developing MI checklists. In addition to its general goals of promoting the use of MI checklists to funders, publishers and the community, and of promoting the development of checklist-supportive tools and databases, the project has two specific goals, with two corresponding sections on the website:

The MIBBI Portal (http://mibbi.org/index.php/MIBBI_portal) exists as an organising point for the many MI checklist projects underway in the community. As a 'shop window' for such projects, the MIBBI Portal increases the likelihood that users and potential contributors will discover a project appropriate to their needs. It also acts as a rallying point for new projects, bringing them into the 'checklist community', encouraging collaboration.

The MIBBI Foundry (http://www.mibbi.org/index.php/MIBBI_foundry) takes the overlapping, hard-to-integrate contents of the various Portal checklists and reworks them into a suite of checklist modules that can be reassembled to fit any number of workflows. A browse interface (http://mibbi.org/index.php/MICheckout) allows users to specify their areas of interest and download compiled sets of these modules, as HTML (for browsing), as XML

Schema (for model-driven software, *inter alia*), as a tab-delimited file (simple spreadsheet), or as a configuration for ISAcreator (http://isatab.sourceforge.net/), for data collection/management/submission.

Ultimately, we envision a situation where different communities 'own' parts of the MIBBI Foundry's content, with shared ownership over common features such as 'organism', 'person', or 'environment'. This body of guidance then becomes a unified expression of the information required when reporting bioscience research, which should then be referenced by software and database developers, viewing it as their driving 'use case'.

THE OBJECTIONS TO FULLER REPORTING

The potential benefits of the widespread sharing of well-annotated data sets are significant; the extraction of further value from existing data; the re-use of data in meta studies; the maintenance of a solid evidence base for claims made in the literature; and so on. However, this is not the whole picture. For those who generate the data, whose worth is assessed through the value they extract from it while they have sole access, and who must invest the time to encode, describe and share it, things can look rather different. Although most are governed by funding conditions that mandate data sharing, such strictures rarely obtain greater than the minimum of effort; 'data generators' must be positively encouraged to share their data.

The arguments against data sharing are of four kinds: (1) that the loss of intellectual property reduces the chances for researchers to maximally exploit their data; (2) that neither the money nor the tools are available to support data sharing; (3) that other's shared data will potentially be of low quality, a problem exacerbated by the difficulty of *post hoc* quality assessment; (4) that others (especially bioinformaticians) will benefit from the investment in the shared data without credit accruing to the originator. However, all of these issues can or have been addressed.

Under the policy of most funders, intellectual property is protected for a period of years after data are generated, allowing for most value to be extracted before the data are shared with the wider community. It is also fairly normal these days to apply for funds for data management as part of a funding application; and as the standards that will support data sharing settle down, so we have seen tools begin to emerge. The issue of data quality is a difficult one. Many funders require that all data are shared, including data that were not the subject of a publication, and therefore have not been peer-reviewed. The solution will most likely come from databases, who will (and in some cases already do [6]) use metadata and statistical analyses to assess the likely quality of a data set, but even data of moderate quality can be useful if suitably annotated (data processing techniques may improve, for example).

The objection that no credit accrues from data reuse is challenging. To acknowledge one's peers is central to the scientific process, yet the re-use of downloaded data is often unacknowledged. Even where it is acknowledged, there is normally no way for standard assessments of academic performance to count the 'impact' of such a thing. The solution is firstly to make data sets 'citeable' through the use of Digital Object Identifiers (DOIs), secondly to ensure that referees are as ardent about missing citations for data sets used as for papers, and lastly to ensure that literature databases, funders and faculties count these as the equivalent of normal citations [7]. A side benefit of making data sets citeable is that the quality of annotation may improve further: if the re-use of a data set is governed by its utility, then it follows that higher quality annotation, by improving utility (both because the data set is better characterised, and because the additional annotation may increase the chance of orthogonal re-use) will cause the data set to be re-used more frequently, with more credit accruing to the generator of those data.

SUMMARY

The MIBBI Project seeks to coalesce fragmented efforts to develop MI checklists, and ultimately to provide a unified set of checklist modules by re-using the work of individual communities. This parallels unification efforts for other kinds of standard (formats, vocabularies) and complements the desire of funders and publishers for increased data sharing. However, the success of the MIBBI Project, and of efforts to increase data sharing generally, depend on the instigation of a system to uniquely identify data sets, on the use of that system by those who assess the worth of researchers, and on the availability of appropriate, robust tools.

REFERENCES

[1]   Smith, B., Ashburner, M., Rosse, C., Bard, J., Bug, W., Ceusters, W., Goldberg, L.J., Eilbeck, K., Ireland, A., Mungall, C.J., OBI Consortium, Leontis, N., Rocca-Serra, P., Ruttenberg, A., Sansone, S.A., Scheuermann, R.H., Shah, N., Whetzel, P.L., Lewis, S. (2007) The OBO Foundry: coordinated evolution of ontologies to support biomedical data integration. *Nat. Biotechnol.* **25**(11):1251–5. doi: http://dx.doi.org/10.1038/nbt1346.

[2]   Jones, A.R., Miller, M., Aebersold, R., Apweiler, R., Ball, C.A., Brazma, A., Degreef, J., Hardy, N., Hermjakob, H., Hubbard, S.J., Hussey, P., Igra, M., Jenkins, H., Julian, R.K. Jr, Laursen, K., Oliver, S.G., Paton, N.W., Sansone, S.A., Sarkans, U., Stoeckert, C.J. Jr, Taylor, C.F., Whetzel, P.L., White, J.A., Spellman, P., Pizarro, A. (2007) The Functional Genomics Experiment model (FuGE): an extensible framework for standards in functional genomics. *Nat. Biotechnol.* **25**(10):1127–33. doi: http://dx.doi.org/10.1038/nbt1347.

[3]   Sansone, S.A., Rocca-Serra, P., Brandizi, M., Brazma, A., Field, D., Fostel, J., Garrow, A.G., Gilbert, J., Goodsaid, F., Hardy, N., Jones, P., Lister, A., Miller, M., Morrison, N., Rayner, T., Sklyar, N., Taylor, C., Tong, W., Warner, G., Wiemann, S., Members of the RSBI Working Group. (2008) The first RSBI (ISA-TAB) workshop: "can a simple format work for complex studies?". *OMICS* **12**(2):143 – 9. doi: http://dx.doi.org/10.1089/omi.2008.0019.

[4]   Taylor C.F. (2007) Standards for reporting bioscience data: a forward look. *Drug Discov Today* **12**(13 – 14):527 – 33. doi: http://dx.doi.org/10.1016/j.drudis.2007.05.006.

[5]   Field, D., Sansone, S.A., Collis, A., Booth, T., Dukes, P., Gregurick, S.K., Kennedy, K., Kolar, P., Kolker, E., Maxon, M., Millard, S., Mugabushaka, A.M., Perrin, N., Remacle, J.E., Remington, K., Rocca-Serra, P., Taylor, C.F., Thorley, M., Tiwari, B., Wilbanks, J. (2009) Megascience. 'Omics data sharing. *Science* **326**(5950):234 – 6. doi: http://dx.doi.org/10.1126/science.1180598.

[6]   Kapushesky, M., Emam, I., Holloway, E., Kurnosov, P., Zorin, A., Malone, J., Rustici, G., Williams, E., Parkinson, H., Brazma, A. (2010) Gene expression atlas at the European bioinformatics institute. *Nucleic Acids Res.* 38(Database issue): D 690 – 8.

[7]   Thorisson, G.A. (2009) Accreditation and attribution in data sharing. *Nat. Biotechnol.* **27**(11):984 – 5. doi: http://dx.doi.org/10.1038/nbt1109-984b.

Guidelines for Reporting of Biocatalytic Reactions

Lucia Gardossi[1,*], Peter J. Halling[2]

[1]Dipartimento di Scienze Farmaceutiche, Università degli Studi di Trieste,
Trieste, Italy

[2]WestCHEM, Department of Pure & Applied Chemistry, University of Strathclyde,
Glasgow G1 1XL, Scotland, U.K.

E-Mail: *gardossi@units.it

Received: 12th January 2010 / Published: 14th September 2010

Abstract

Applied biocatalysis is the general term for the transformation of natural and non-natural compounds by enzymes for preparative purposes.
Because of this, the term biocatalysis is also used to refer to the
application of enzymes in chemistry. There is a steadily rising number
of publications reporting the use of biocatalysis. Unfortunately, the
value of many of these publications is limited because essential information about the experiments is not presented. Recently, the scientific
committee of the European Federation of Biotechnology Section on
Applied Biocatalysis (ESAB), taking also inspiration from the
STRENDA guidelines, prepared and published guidelines for the correct reporting of experiments in biocatalysis. The present manuscript
would like to draw attention to some specific relevant experimental
issues, which differentiate applied biocatalysis from fundamental
enzymology and deserve particular methodological consideration.

Introduction

A wide variety of complex molecules are accepted by enzymes, including synthetic molecules with structures very different from the substrates found in nature. The surge in
practical utilization of biocatalysts is driven by their versatility, regio-, chemo-, and

enantio-selectivity, along with the necessity for the chemical industry to move to environmentally compatible catalysts and processes [1, 2]. Biocatalysis already represents an important tool for the production of fine chemicals and especially pharmaceuticals.

As a discipline, applied biocatalysis touches upon different fields. Overall, the development of a complete biocatalytic process for practical industrial applications involves the contributions of disciplines as diverse as molecular biology, enzymology, microbiology, biotechnology, organic chemistry, materials chemistry and chemical engineering. Gaining a comprehensive and detailed knowledge of all aspects of biocatalyst behaviour is a very difficult proposition.

Recently, the scientific committee of the European Federation of Biotechnology Section on Applied Biocatalysis (ESAB) approved a document, which provides practical and schematic guidelines for reporting experimental data effectively while, most importantly, enabling other scientists to reproduce the experiments. The aim of standardized reporting is also to create value for many other scientific disciplines, especially where biocatalysis must be integrated in multi-step syntheses.

The ESAB guidelines take the STRENDA checklist as a starting point, trying to avoid an unnecessary duplication of standards and checklists in overlapping scientific areas. The resulting document [3] incorporates the STRENDA checklist (http://www.strenda.org/), and includes a list of some items specific to applied biocatalysis and explanations for those who are less familiar with the field.

Therefore, reports of experiments in applied biocatalysis should follow the STRENDA checklist, although not all items on the checklists will need to be specified for every study in biocatalysis, but they rather act to prompt for information that will be required where necessary for the type of experiments reported or that will be useful to supply if available.

Among the topics discussed in the ESAB guidelines, here we would like to put the emphasis on those issues that correspond to essential methodological differences between applied biocatalysis and fundamental enzymology. They can be schematically referred to four major conceptual differences: 1) In applied biocatalysis the emphasis is most often towards the preparative application of enzymes rather than on their full characterization; 2) Since the main priority is to make the enzyme/biocatalyst suitable for preparative applications, the identification of optimal experimental conditions as well as the possibility to make a clear comparison between different sets of conditions are generally of primary importance; 3) Biocatalysts most often are used under "non-physiological" – sometimes extreme – experimental conditions; 4) The audience of applied biocatalysis is multi-disciplinary, constituted not only by enzymologists or biochemists, so that the attention is focused on widely varying experimental issues. While fundamental scientists will be more interested in understanding the catalyst behaviour, synthetic chemists will be more focused on the reproducibility and efficiency of the synthetic protocol, whereas process engineers will be concerned with the scalability of the process or mass transfer issues. The following paragraphs would like to provide a brief overview on the complexity of the experimental systems which are the object of study in applied biocatalysis, underlining the crucial concepts which must be taken into account in the reporting of experimental data. Most of these concepts have been developed thanks to fundamental researches carried out in the last two decades, and they enabled applied biocatalysis to abandon purely empirical investigative strategies, thus becoming a more rigorous discipline. Nevertheless, it emerges clearly that the complete understanding of enzyme behaviour in such complex systems still represents a formidable task.

DESCRIPTION OF EXPERIMENTS IN APPLIED BIOCATALYSIS: CORRECT IDENTIFICATION OF THE BIOCATALYST

The development of fermentation processes and biochemical methods specifically aimed at the production of enzymes make it possible to manufacture enzymes as purified, well-characterized preparations even on a large scale. The use of recombinant gene technology has further improved manufacturing processes and enabled the commercialization of enzymes that could previously not be produced on large scale. Furthermore, modern biotechnologies, such as protein engineering and directed evolution, have further revolutionized the development of industrial enzymes. These advances have made it possible to provide tailor-made enzymes displaying new activities and adapted to new process conditions, enabling a further expansion of their industrial use [4, 5].

As mentioned above, most often in applied biocatalysis the attention is focused on the identification of the optimal experimental conditions that allow a biocatalyst to perform an efficient biotransformation for preparative purposes. Therefore, biocatalysts may often be less well characterised than is usual in pure academic enzymology. The only requirement, which is equally important for biocatalysis, is that the catalysts used must be specified unambiguously.

Commercial enzyme products need to be described by citing the manufacturer with name and address, complete product name, code from manufacturer, date of sample or batch number.

As the general industrial aim is to minimize the enzyme purification and operate the process with the crudest form of catalyst possible (usually a lysate), most studies are made with crude or partially purified enzymes. As a matter of fact, many commercial enzymatic preparations contain a quite low percentage of protein (sometimes < 1%), whereas a considerable amount of stabilizing agents, mostly polyols and salts, are present. The detailed description of the impurities or additives is of primary importance for the reproducibility of the experiments. It must be underlined that the presence of impurities and additives in the original enzymatic preparation must also affect the immobilization protocols (see below).

Possible interference with enzymatic assays or protein determination procedures must be carefully evaluated. This is especially crucial when different enzymes or experimental conditions are compared in terms of efficiency. It is worth noting that many academic enzyme catalysis studies are carried out using pure enzymes and their results might not always be easily translatable at industrial scale.

REPRODUCIBILITY OF EXPERIMENTS AND IDENTIFICATION OF THE BIOCATALYST: THE CASE OF IMMOBILIZED BIOCATALYSTS

Biocatalysts can be used as microbial or plant cells or as isolated enzymes. However, enzymes usable for a given reaction are often hampered by lack of long-term stability under process conditions, and also by difficulties in recovery and recycling. These problems can be overcome by immobilizing the enzymes on solid supports, so that the biocatalysts are used as insoluble particles.

Immoblization may provide the following advantages:

- repeated or continuous use,
- easy separation from the reaction mixture,
- enhanced stability,
- possible modulation of the catalytic properties,
- prevention of protein contamination in the product,
- easier prevention of microbial contaminations.

Since the first uses of biocatalysts in organic synthesis dating back almost a century, researchers have tried to identify methods for linking an enzyme to a carrier [6].

A single broadly applicable method for enzyme immobilization still needs to be discovered. The most frequently used immobilization techniques fall into four categories:

- non-covalent adsorption or deposition,
- covalent attachment (mostly used for isolated enzymes),
- entrapment in a polymeric gel (mostly used for whole cells), membrane or capsule,
- cross-linking of an enzyme.

All these approaches are a compromise between maintaining high catalytic activity while achieving the advantages of immobilization.

Support binding can simply exploit weak hydrophobic and van der Waals interactions, or stronger ones such as ionic. More appropriate for industrial applications is the covalent binding of the enzyme to the support since it has the advantage that the enzyme cannot be leached from the solid support.

epoxy support

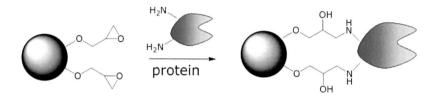

Figure 1. An example of covalent immobilization of enzymes on a functionalized solid support: amino groups of lysine side chains are used for the nucleophilic opening of the epoxy rings present on the surface of the support.

Enzymes can be also entrapped in polymer networks such as an organic polymer or a silica sol-gel, or a membrane device such as hollow fibres or a microcapsule. The physical restraints generally are too weak, however, to prevent enzyme leakage entirely. Hence, additional covalent attachment is often required.

Carrier-free immobilized enzymes are prepared by the cross-linking of enzyme aggregates or crystals, using a bifunctional reagent. This procedure leads to macroparticles, such as cross-linked enzyme crystals (CLECs) and cross-linked enzyme aggregates (CLEAs). Entrapment is more suited for the immobilization of whole cells. The increasing knowledge of enzyme structures and mechanism has enabled more controlled immobilizations. Information derived from protein sequences, 3D-structures, and reaction mechanism are combined with the properties of carriers (functional groups, hydrophobicity, magnetic properties) and physical/chemical methods in order to develop optimal immobilization strategies on a rational basis. As a matter of fact, a remarkable number of scientific publications describe protocols for developing efficient immobilized enzymes [7]. In these studies, a clear protocol of immobilization with full characterization of the biocatalyst should be reported, by specifying a) amount of support and enzyme used; b) amount of the enzyme that was actually bound to the support, (*e. g.* by measuring the free protein present in solution before and after immobilization); c) the activity of the immobilised preparation; d) the residual water content (to allow an estimation of the activity on a dry weight basis); e) data on the physical-chemical properties of the support (*e. g.* chemical nature, porosity, surface area, size of the particles, type of functional groups present on the surface and their density, when available); f) any further relevant information necessary to reproduce and compare different protocols, such as the distribution of the enzyme within particles (when feasible) [8] or the diffusion of substrates and products between the pores of the particles and the bulk reaction medium, when mass transfer limitations are relevant.

The use of immobilized enzymes may imply that the enzymatic activity is measured under conditions of strong mass transfer limitation. This must be taken into account particularly in stability studies, since in that case the intrinsic activity of the catalyst can fall substantially with little change in measured activity – because this is still largely limited by the same mass transfer rate [9, 10].

Since the ultimate objective of enzyme immobilization is to obtain a re-cyclable biocatalyst, the evaluation of operational stability (i.e. under the conditions of the useful biocatalytic synthesis) is of major importance. This can be studied in a continuous reactor, or by recycling the catalyst in repeated batches [1]. In the latter case, the treatment (e.g. rinsing) of the biocatalyst after recovery can play a crucial role, so these procedures should be specified. Presentation of results must clearly distinguish total time spent under reaction conditions from elapsed time, particularly when the cycle includes an extended storage time between successive batch reactions.

REPRODUCIBILITY OF EXPERIMENTS CARRIED OUT UNDER NON-PHYSIOLOGICAL CONDITIONS: "NON CONVENTIONAL MEDIA"

The economic feasibility of a biocatalytic process at industrial level depends on several factors. The usual requirement is to achieve product concentrations comparable to chemical processes, namely at least 50 – 100 g/l. In nature, enzymes usually work at millimolar levels of substrate so that such high concentrations are achievable only thanks to proper process development, as well as protein engineering allowing the enzyme to maintain sufficient activity. Most biocatalytic industrial processes still operate in aqueous environments that generally correspond to low product concentrations because of the poor solubility of most organic molecules in water.

For a long period it was thought that enzymes should be restricted to their natural environment: diluted aqueous reaction media at ambient pressure and temperatures. Indeed industrially enzymes were first employed only for hydrolytic processes. With an increase of the range of enzyme applications the aqueous medium became limiting. These observations led to the introduction of the so called "non-conventional media" in biocatalysis. By definition, a non-conventional medium is any system different from a dilute aqueous solution of an enzyme.

Among the "non conventional systems" those employing organic solvents are the most widely used and can be classified into three different categories:

- Enzyme suspended in a monophasic organic solution

- Monophasic aqueous/organic solution

- Biphasic aqueous/organic solution.

The first examples of biocatalysis in organic solvents actually date back to before 1900, and in the 1930s Ernest Alexander Sym published ground breaking work on the activity of pancreatic-lipase preparations in organic solvents, finding a correlation between the equilibrium position and the water concentration of the system [11, 12]. However, biocatalysis in organic solvents did not "take off" until the 1980s when the application of enzymes in monophasic organic solvents was finally studied in a systematic way [13].

These systems can be obtained by replacing the bulk water by a water immiscible organic solvent and this leads to a suspension of the solid enzyme in a monophasic organic solution. Although the biocatalyst seems to be dry from a macroscopic view, it must have necessary residual bound water to remain catalytically active. As a consequence, these systems are also referred to as "low-water media".

Indeed, a "low-water medium" can be constituted simply by the neat organic substrates, without the extra addition of solvents. As an example, immobilized lipases suspended in a mixture of triglycerides and alcohols catalyze the transesterification that produces biodiesel (alkylesters of fatty acids) and glycerol [14]. The same immobilized lipases are used also on industrial scale for the synthesis of polyesters via polycondensation: the biocatalyst is employed at temperatures above 60 °C and suspended in highly viscous mixture composed of the neat substrate monomers [15].

An essential understanding of the behaviour of enzymes in organic solvent came from the studies on the effect of water activity (a_w), a crucial parameter for determining the correct degree of hydration of enzymes in non-aqueous media [16]. The concept of water activity can be related to the "available water" present in the system, namely the water which is "free" to react, hydrate other molecules or partition in other phases. The "availability" of the water will be lower, for instance, in the presence of polar protic solvents, which are prone to establish hydrogen bonds with water molecules thus reducing their "freedom". When a system reaches the equilibrium, the water activity will be the same in all phases. Therefore, the reaction, the enzyme hydration and ultimately the enzyme activity will be affected by the a_w rather than by the overall water concentration in the system.

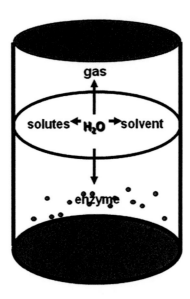

Figure 2. A schematic representation of the distribution of water among phases in a system employing enzymes suspended in mono-phasic organic medium. The hydration of the biocatalyst will depend on the amount of "available water" (i. e. water activity) rather then on the total amount of water present in the system. Since at equilibrium the water activity will be the same in all phases, it can be measured in the most accessible one, generally the gas phase *via* the measurement of vapour pressure of water.

To perform at its best each enzyme requires an optimal degree of hydration that is guaranteed by working at controlled water activity (a_w). Many lipases work efficiently at rather low values of water activities (<0.2) whereas most other enzymes such as proteases require water activities values above $0.4 - 0.5$.

The water activity can be determined by measuring the vapour pressure in the gas phase of the system which has reached equilibrium. A hygrometer can be used for this purpose, although most of the instruments available on the market are easily damaged by organic solvents. Alternative options are: a) to dry or bring to a defined water content all the reaction components; b) to equilibrate defined parts of the reaction mixture to the desired value of water activity by making use, for instance, of couples of hydrated salts [17]. In general, enzymes work well in relatively non-polar organic solvents (log P > 2), those that do not mix with water and therefore do not compete for the "free water" nor interfere with the hydrogen bonds responsible for the native active conformation of the enzyme.

Monophasic aqueous/organic solutions are usually employed for the transformation of lipophilic substrates which are poorly soluble in an aqueous system and which would therefore react at a low reaction rate. These systems consist of water and a water-miscible organic co-solvent such as dimethyl sulfoxide, dimethyl formamide, tetrahydrofuran, dioxane, acetone or a short chain alcohol. If the proportion of the solvent exceeds a certain threshold (which depends on the enzyme and the co-solvent used), the enzyme undergoes denaturation.

When dealing with these "non-physiological" reaction media, some further methodological precautions must be taken in the reporting of experimental conditions. For instance, when the medium is not a dilute aqueous solution, to get a true pH, the electrode needs to be calibrated with pH standards in the same medium (if suitable standards are known). More commonly, the electrode will have been calibrated as normal using dilute aqueous standards, in which case this should be stated clearly and the readings referred to as 'apparent pH'. Sometimes pH will be set or read before the medium is in the final form used for reaction (*e. g.* before heating or adding a co-solvent). Again this must be specified.

During the last decades much attention has been paid, by both academia and industry, to the development of further new solvents, possibly environmentally friendly. Two new classes of solvents are the most cited in the literature: ionic liquids (IL) and supercritical fluids (SF) [18]. Ionic liquids usually consist of an organic cation, often containing a nitrogen heterocycle, and an inorganic anion. First assays of biocatalysed reactions in these unusual media were remarkably successful showing that enzymes not only tolerate these solvents, but, indeed, that they are also stable and the activity is comparable or even better than in organic solvents [19]. However, so far most studies are of an exploratory nature. The relationships between structure of the IL and activity or stability of the enzyme is not yet clearly understood. This is, however, a prerequisite for this methodology to be developed to the full. Similarly to their behaviour in organic solvents, enzymes in ionic liquids require a certain degree of hydration. This should be guaranteed by controlling the water activity (a_w) of the system. This is particularly difficult in the case of the hygroscopic ionic liquids that, depending on the synthetic procedure, the drying process or the storage conditions, can contain very different amounts of water.

Supercritical fluids are materials above their critical temperature, T_c, and critical pressure, P_c. Properties of supercritical fluids lie between the properties of liquids and gases. The densities of supercritical fluids are comparable to those of liquids, while the viscosities are comparable to those of gases. Supercritical fluids are an environmentally friendly alternative to organic solvents as media for biocatalysis because they can be non flammable (e. g. CO_2) and at the end of enzymatic processes, traces of sc-fluids can be removed by depressurisation.

The combination of ILs with supercritical fluids can be a good strategy to circumvent the use of organic solvents to recover solutes. Thanks to the high solubility of supercritical fluids in the ILs, the mass transfer of solutes is increased, and it is possible to couple (bio)transformations in ILs with extraction by supercritical fluids.

The most used SF for biocatalysed processes is supercritical-CO_2, but also other supercritical fluids like supercritical-ethane have been used successfully. In biocatalytic processes, the gas-like viscosity enhances mass transfer rates of reactants to the active sites of enzymes that are dispersed in the supercritical fluid. In this way reactions that are limited by the rates of diffusion, rather than intrinsic kinetics, will proceed faster in supercritical fluids than in normal liquids. A key feature of biocatalysis in supercritical fluids is the tunability of the medium. Small changes in pressure lead to significant changes in density, thus altering all density-dependent solvent properties (dielectric constant, solubility parameter and partition coefficient).

BIOCATALYZED REACTIONS IN MULTI-PHASE SYSTEMS: CORRECT DESCRIPTION OF THE SYSTEM AND MONITORING OF THE REACTIONS

It must be underlined that in most cases biocatalyzed reaction systems are constituted by separated phases. This can result, for instance, from the employment of immobilized enzymes, enzymes suspended in low-water media, or from the presence of solid particles of substrates or products. In these cases, obtaining meaningful samples for monitoring reaction progress becomes difficult. Although the removal of immobilized enzymes by filtration or centrifugation is a common procedure generally accepted, the most reliable method consists in terminating and extracting the entire reaction mixture, using separate reaction vessels to explore different time-points, since the behaviour of the reaction mixture as a whole will be perturbed by sample removal [20].

If samples of a multi-phase reaction mixture are withdrawn under vigorous agitation, this always requires careful checks that the samples really are representative, with determination of the relative volumes or masses of different phases by a suitable approach. When a sample of one liquid phase can be removed uncontaminated by others (*e. g.* after briefly stopping agitation), analysis should give reproducible and meaningful concentrations. However, samples removed according to this procedure will not give complete information about the reaction progress, particularly for compounds mainly distributed into other phases. Furthermore, the total volume removed for analysis must account only for a small fraction of the total volume of this liquid phase.

The procedure for mixing the reaction system must also be described in detail, because in multi-phase systems the apparent reaction rates are often determined by mass transfer limitations. In relation to this, the diameter and height of the vessels are also parameters affecting the reaction progress and they must be specified.

Some biocatalyzed methods involve the use of enzymes in multiphase systems and under rather "extreme" conditions. An example is enzymatic synthesis carried out in reaction mixtures with mainly undissolved substrates and/or products. In this synthetic strategy the compounds are present mostly as pure solids [21]. Although these reaction mixtures usually consist largely of solids, it has been recognized that a liquid phase is essential for enzymatic activity. In a reaction with two solid substrates, this usually means the addition of a solvent (sometimes referred to as "adjuvant") to the mixture. One of the two substrates can be a liquid at the reaction temperature, so that it can then be used as the "solvent" to partially dissolve the other substrate. In some cases a liquid phase can be formed from two solid substrates by eutectic melting, when the reaction temperature lies below the melting points of the pure substrates, but above their eutectic temperature. The physical appearance of such reaction mixtures can vary widely depending on the ratio of the different components and on the nature of the liquid phase used. Thus, there are mainly solid systems or dilute suspensions in a large liquid phase in which a product can precipitate because its solubility in the solvent used is extremely low. When product precipitates the reaction yields are improved so that the necessity to use organic solvents to shift the thermodynamic equilibrium toward synthesis is reduced and synthesis is made favourable even in water. Although substrates are usually largely undissolved in such systems, very high conversion yields were observed in many of the reactions studied in the literature. The thermodynamics of these reaction systems have been investigated, resulting in methods to predict the direction of a typical reaction a priori. Furthermore, studies on kinetics, enzyme concentration, pH/temperature effects, mixing and solvent selection have opened new perspectives for the understanding, modelling, optimisation and the possible large scale application of such a strategy [22]. It is clear that several aspects of reaction systems with suspended substrates are significantly different from those in solution. The presence of solid substrates has important consequences for the reaction kinetics and thermodynamics and it requires different strategies for reaction engineering. The majority of the published work on this type of reaction was related to the synthesis of peptides, but the synthesis of beta-lactam antibiotics, glycosides, glycamides, esters and polyesters has also been reported.

Enzymes are able to recognize and transform substrates even when these molecules are anchored on solid supports [23]. When employed for solid phase synthesis, enzymes are generally dissolved in an aqueous buffer and react with the substrate anchored on a water-insoluble resin. The substrate is generally separated from the resin by a chemical linker that must be cleaved in selective and mild conditions at the end of the transformation in order to recover the product.

Figure 3. Enzymes, dissolved in buffer, can accept substrates anchored on solid supports. Enzymes can be used either for chemical transformation of target molecules or for the selective cleavage of linkers.

Examples of applications of enzymes on immobilized substrates are chemo-enzymatic synthesis of compound libraries "on-bead", peptide synthesis, screening for enzyme substrates or inhibitors in combinatorial libraries, applications in micro-array technologies and enzymatic optical resolution.

Many hydrolytic enzymes (proteases, esterases, glycosidases and amidases) have also been investigated for their ability to selectively cleave enzyme-scissile linker groups.

Evaluation of libraries of compounds generated by combinatorial chemistry has been appreciated during the last decade as an efficient and rapid approach to synthesise and screen arrays of compounds on a nanoscale. The application of enzymes in combinatorial chemistry has attracted significant attention and enzymatic methods have opened up advantageous alternatives to classical chemical techniques, since enzyme-catalysed transformations often proceed under very mild conditions and are highly selective. The ability of the enzyme to catalyse reactions on solid phase strongly depends on the dimension of the protein compared to the pore size of the resin. The permeability of enzymes into the resin can be improved in two ways: *i*) by creating in the polymer porosities of such dimensions that firstly the enzyme can freely approach the linked substrate and then undergo the conformational modification necessary to recognise and transform it; and *ii*) by inducing enlargement of the cavities inside the resin through efficient solvation and swelling of the polymer in the solvent, that in most cases consists of a buffered aqueous solution, due to the necessity of dissolving the molecules of enzyme.

The use of enzyme in solid phase has found application also in microarray technology. This technology has found a promising route in the use of biocatalysts for the development of highly selective assays under mild operating conditions.

The final example of a "non-conventional" multi-phase enzymatic system is represented by solid-gas biocatalysis, a promising technology for the development of new clean industrial processes. The use of enzymes or whole cells at the solid-gas interface offers some very interesting features since total thermodynamic control of the system can easily be achieved

[24]. Solid-gas biocatalysis presents many advantages compared to other systems (mono- or bi-phasic liquids): very high conversion yields compatible with a high production rate at a minimal plant scale, more efficient mass transfer, reduced diffusion limitations due to low gas viscosity and better stability of enzymes and cofactors. Many enzymes have been explored in solid-gas systems such as alcohol oxidase in ethanol oxidation, alcohol dehydrogenase for alcohol and aldehyde production.

TOPICS OF PARTICULAR RELEVANCE FOR APPLIED BIOCATALYSIS: SPECIFICITY AND SELECTIVITY

The application of enzymes is particularly valuable in the fine-chemical sector because of their specificity and selectivity. Enzymes may be making a 'distinction' between two or more possible substrates, or between two or more possible reactions on a single substrate. In any statement about specificity or selectivity, it should be made absolutely clear what comparison is being made. Since the words 'selectivity' and 'specificity' are often used with different meanings, it is recommended to use the prefixes 'chemo-', 'regio-' and 'enantio-' because they do have clearly defined meanings, and their use might enable the comparison to be made clear.

Biocatalysts have a pivotal role in the industrial production of enantiomerically enriched chiral fine chemicals. The increasing size and complexity of fine chemicals (agrochemical, pharmaceuticals) along with the development of new materials (e.g. liquid crystals and polymers) imply that these molecules frequently contain multiple chiral centres. Moreover, due to the fact that the current USA Food and Drug Administration (FDA) regulations demand proof that the non-therapeutic isomer be nonteratogenic, compounds with a chiral centre are usually manufactured in single isomeric form. As chiral molecules, enzymes may be able to discriminate between a pair of enantiomers, or catalyze a chemical transformation resulting in the introduction of a new chiral centre in an enantiopure form [25].

Generally, the optical purity of a chiral compound is expressed by the enantiomeric excess (ee). For asymmetric syntheses generating a new chiral centre, product ee values will normally be constant as the reaction proceeds, and do directly characterise the biocatalyst [26]. However, in reactions involving the resolution of a racemate, where the two enantiomer substrates are bio-transformed at different rates (i.e. kinetic resolutions) the ee value of substrates and products will vary throughout the progress of the reaction [27]. When an enantio-resolution must be described and the enantioselectivity of an enzyme characterized quantitatively, some parameter expressing the differences of activation energies for the two competing reactions should be used.

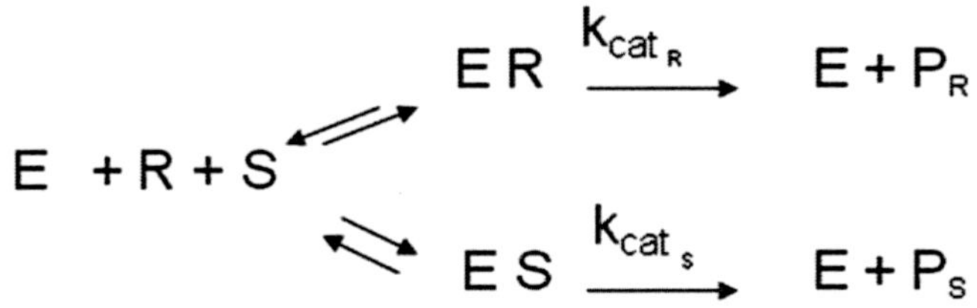

Figure 4. Scheme of the enzymatic resolution of a couple of enantiomers.

In this respect, the enantioselectivity can also refer to a ratio of the specificity constants (k_{cat}/K_m) for the two enantiomers, and can be treated quantitatively, provided it is clear which reactions are being compared [28]. However, since the experimental evaluation of k_{cat}/K_m is quite laborious, generally it is preferred to calculate the "E value", which expresses the ration of specificity constants in terms of relationship between the conversion (c) and the ee of the recovered substrate fraction (ee(S)) or the enantiomeric excess of the product (ee(P)).

$$E = \frac{\ln[(1-c)(1-ee(S))]}{\ln[(1-c)(1+ee(S))]} = \frac{\ln[1-c(1+ee(P))]}{\ln[1-c(1-ee(P))]}$$

Figure 5. Equation used for the experimental calculation of the "E value" (ratio of specificity constants).

Therefore, only the "E value" can sensibly be compared in order to evaluate the performance of the biocatalyst under different conditions [29, 30].

Hydrolases are used in enzymatic resolution because of their ability to preferentially hydrolyse one enantiomer of a racemic substrate, thus providing a means of separation. Hydrolytic enzymes also effectively catalyse enantio-complementary reverse hydrolysis (esterification, transesterification, aminolysis or amidation), providing access to both enantiomers of a desired product. The drawback to the usual strategy of enzymatic resolution is that the desired enantiomer is obtained in a maximal 50% yield, which is too low to allow a positive economic and environmental balance for such transformations. To overcome this limitation different strategies, generally referred to as "deracemization", have been developed that allow the transformation of both enantiomers of a racemate into a single enantiomer of the product. As an example, in situ racemization of substrate combined with kinetic resolution leads to the concept of "dynamic kinetic resolution" (DKR) [31].

Enzymatic reduction of carbonyls is a powerful tool for the production of optically pure chiral alcohols from prochiral compounds. Dehydrogenases can act as asymmetric catalysts and the theoretical yield of a single enantiomer of the chiral alcohol is 100%. Production of chiral alcohols through the asymmetric reduction of prochiral carbonyls has been thoroughly investigated using whole cells of bacteria and yeasts.

CONCLUSIONS

The previous paragraphs highlight the outstanding potential of enzymes as catalysts and the advances of applied biocatalysis towards more and more sophisticated experimental systems. This implies that controlling the properties of the biocatalytic system through rigorous experimental procedures is rather challenging.

Fundamental enzymology is expected to provide an essential support to applied biocatalysis by sustaining the necessary methodological evolution of this discipline and thus promoting the full exploitation of enzyme catalytic potential.

REFERENCES

[1]   Bommarius, A.S. and Riebel, B. (2004) *Biocatalysis* Wiley-VCH, Weinheim, Germany.

[2]   Ghisalba, O., Meyer, H.P., Wohlgemuth, R. (2009) Industrial Biotransformation. In *Encyclopedia of Industrial Biotechnology.* Ed. Flickinger M. C., Wiley, Hoboken, NJ.

[3]   Gardossi, L. Poulsen, P.B., Ballesteros, A., Hult, K., Švedas, V.K., Vasiæ-Raèki, D., Carrea, G., Magnusson, A, Schmid, A., Wohlgemuth, R., J. Halling, P.J. (2010) Guidelines for reporting of biocatalytic reactions. *Trends Biotechnol.* **28**(4):171–180.

[4]   Arnold, F. H. (2001) Combinatorial and computational challenges for biocatalyst design. *Nature* **409**:253 – 257.
       doi: http://dx.doi.org/10.1038/35051731.

[5]   Leisola, M., Turunen, O. Protein engineering: opportunities and challenges. *Appl. Microbiol. Biotechnol.* **75**:1225 – 1232.
       doi: http://dx.doi.org/10.1007/s00253-007-0964-2.

[6]   The EFB Working Party on Immobilised Biocatalysts (1983) Guidelines for the characterization of immobilised biocatalysts. *Enzyme Mic. Tech.* **5**:304 – 307.
       doi: http://dx.doi.org/10.1016/0141-0229(83)90082-0.

[7]   van Roon, J.L. Groenendijk, E., Kieft, H., Schroën, C.G.P.H., Tramper, J., Beeftink, H.H. (2005) Novel approach to quantify immobilized-enzyme distributions. *Biotechnol. Bioeng.* **89**:660 – 669.
       doi: http://dx.doi.org/10.1002/bit.20345.

[8]   Sheldon, R.A., (2007) Enzyme Immobilization: The Quest for Optimum Performance. *Adv. Synth. Catal.* **349**:1289 – 1307.
       doi: http://dx.doi.org/10.1002/adsc.200700082.

[9]   Hateful, U., Gardossi, L., Magner, E. (2009) Understanding enzyme immobilization. *Chem. Soc. Rev.* **38**:453 – 468. doi: http://dx.doi.org/10.1039/b711564b.

[10]   Roon, J.L., Arntz, M.M.H.D., Kallenberg, A.I., Paasman, M.A., Tramper, J., Schroën C.G.P.H., Beeftink, H.H. (2006) A multicomponent reaction-diffusion model of a heterogeneously distributed immobilized enzyme. *Appl. Microbiol. Biotechnol.* **72**:263 – 278. doi: http://dx.doi.org/10.1007/s00253-005-0247-8.

[11]   Sym, E. A. (1933) *Biochem. Z.* **258**:304 – 324.

[12]   Sym, E. A. (1936) *Enzymologia* **1**:156 – 160.

[13]   Klibanov, A.M. (2001) Improving enzymes by using them in organic solvents. *Nature* **409**:241 – 246. doi: http://dx.doi.org/10.1038/35051719.

[14]   Salis, A., Pinna, M., Monduzzi, M., Solinas, V. (2005) Biodiesel production from triolein and short chain alcohols through biocatalysis. *J. Biotechnol.* **119**:291 – 299.

[15]   Binns, F., Harffey, P., Roberts, S.M., Taylor, A. (1999) Studies leading to the large scale synthesis of polyesters using enzymes. *J. Chem. Soc. Perkin Trans.* **1**:2671 – 2676. doi: http://dx.doi.org/10.1039/a904889h.

[16]   Halling, P.J. (2000) Biocatalysis in low-water media: understanding effects of reaction conditions. *Current Opinion in Chemical Biology* **4**:74 – 80. doi: http://dx.doi.org/10.1016/S1367-5931(99)00055-1.

[17]   Bell, G., Halling, P.J., May, L., Moore, B.D., Robb, D.A., Uljin, R., Valivety, R.H. et al. (2001) Methods for measurement and control of water in non- aqueous biocatalysis. In *Methods in Biotechnology: Enzymes in Nonaqueous Solvents* (E.N. Vulfson, P.J. Halling and H.L. Holland eds.) pp 105 – 126. Totowa, NJ, U.S.A., Humana Press. doi: http://dx.doi.org/10.1385/1-59259-112-4:105.

[18]   Cantone, S., Hanefeld, U., Basso, A. (2007) Biocatalysis in non-conventional media- ionic liquids, supercritical fluids and the gas phase. *Green Chemistry* **9**:954 – 971. doi: http://dx.doi.org/10.1039/b618893a.

[19]   van Rantwijk, Madeira Lau, R., Sheldon, R.A. (2003) Biocatalytic transformations in ionic liquids *Trends Biotechnol.* **21**:131 – 138. doi: http://dx.doi.org/10.1016/S0167-7799(03)00008-8.

[20]   van Roon, J.L., Schroën, C.G.P.H., Tramper J., Beeftink, H.H. (2007) Biocatalysts: Measurement, modelling and design of heterogeneity. *Biotechnol. Adv.* **25**:137 – 147. doi: http://dx.doi.org/10.1016/j.biotechadv.2006.11.005.

[21]   Ulijn, R.V., De Martin, L., Gardossi, L., Halling, P.J. (2003) Biocatalysis in reaction mixtures with undissolved solid substrates and products. *Curr. Org. Chem.* **7**:1333 – 1346.
doi: http://dx.doi.org/10.2174/1385272033486422.

[22]   Ulijn, R.V., De Martin, L., Halling, P.J., Janssen, A.E.M., Gardossi, L., Moore, B.D. (2002) Solvent selection for solid-to-solid synthesis *Biotech. Bioeng.* **80**:509 – 515.
doi: http://dx.doi.org/10.1002/bit.10396.

[23]   Halling, P.J., Ulijn, R.V., Flitsch, S.L. (2005) Understanding enzyme action on immobilised substrates. *Current Opinion in Biotechnology* **16**:385 – 392.
doi: http://dx.doi.org/10.1016/j.copbio.2005.06.006.

[24]   Lamare, S., Lamare, S., Legoy, M.D., Graber, M. (2004) Solid/gas bioreactors: powerful tools for fundamental research and efficient technology for industrial applications. *Green Chem.* **6**:445 – 458.
doi: http://dx.doi.org/10.1039/b405869k.

[25]   Patel, R.N. (2006) Synthesis of chiral intermediates for pharmaceuticals. *Current Organic Chemistry* **10**:1289 – 1321.
doi: http://dx.doi.org/10.2174/138527206777698011.

[26]   Nakamura, K, Matsuda, T. (2006) Biocatalytic reduction of carbonyl groups. *Current Organic Chemistry* **10**:1217 – 1246.
doi: http://dx.doi.org/10.2174/138527206777698020.

[27]   Trost, B.M. (2004) Asymmetric catalysis: an enabling science. *Proc. Natl. Acad. Sci. U.S.A.* **101**:5348 – 5355.
doi: http://dx.doi.org/10.1073/pnas.0306715101.

[28]   Cornish-Bowden, A. (2004) "Fundamentals of enzyme kinetics", 3rd ed, Portland Press, London.

[29]   Chen, C.-S., Fujimoto, Y., Girdaukas, G., Sih, C.J. (1982) Quantitative analysis of biochemical kinetic resolutions of enantiomers. *J. Amer. Chem. Soc.* **104**:7294 – 7299.
doi: http://dx.doi.org/10.1021/ja00389a064.

[30]   Straathof, A.J.J., Rakels, J.L.L., Heijnen, J.J. (1992) Kinetics of the enzymatic resolution of racemic compounds in bi-bi reactions. *Biocatalysis* **7**:13 – 27.
doi: http://dx.doi.org/10.3109/10242429209003658.

[31] Gadler, P., Glueck, S.M., Kroutil, W., Nestl, B.M., Larslegger-Schnell, B., Uer-bacher, B.T., Wallner, S.R., Faber, K. (2006) Biocatalytic approaches for the quantitative production of single stereoisomers from racemates. *Biochemical Society Transactions* **34**:296–300.
doi: http://dx.doi.org/10.1042/BST20060296.

SABIO-RK: Kinetic Data for Reaction Mechanism Steps

Ulrike Wittig*, Andreas Weidemann, Heidrun Sauer-Danzwith, Sylvestre Kengne, Isabel Rojas and Wolfgang Müller

Scientific Databases and Visualization Group, EML Research gGmbH/
Heidelberg Institute for Theoretical Studies (HITS),
Schloss-Wolfsbrunnenweg 33, 69118 Heidelberg, Germany

E-Mail: *Ulrike.Wittig@eml-r.villa-bosch.de

Received: 20th January 2010 / Published: 14th September 2010

Abstract

SABIO-RK is a curated database containing kinetic information not only for biochemical reactions but also for individual steps of the reaction mechanism manually extracted from literature. Data in SABIO-RK comprises information about reaction participants including enzyme properties, biological locations (organism, tissue etc.), kinetic parameters and rate equations determined for the reaction, and the experimental conditions used for the determination. To understand biochemical reactions and their kinetic behaviour not only reaction details and kinetic properties of the biochemical reactions but also details of the reaction mechanism are essential. To meet these requirements additionally to the kinetic data of the reactions SABIO-RK offers a graphical representation of the mechanism of a selected reaction as a survey and also a detailed listing of all individual reaction steps as separate entries.

Introduction

SABIO-RK (http://sabio.villa-bosch.de/) [1, 2] is a database containing information about biochemical reactions and their kinetic properties based on literature information and data from automatic submissions from lab experiments (publication in progress). The kinetic data are related to the organism, tissue or cell type and to the environmental conditions in which they were determined. Beside the information about the overall reactions SABIO-RK offers new feature information about individual mechanism steps.

Most of the available pathway and enzyme databases (KEGG [3], BRENDA [4], IUBMB [5], IntEnz [6], Rhea [7], Reactome [8] etc.) contain information about enzymes and biochemical reactions only focussing on the overall biochemical reactions catalysed by enzymes. This does not allow a detailed analysis of the reaction mechanism by searching for elementary steps of the overall reaction. The MACiE database [9] offers stepwise information about the reaction mechanism based on structural information of the proteins and chemical compounds. For each EC sub-subclass where there is a crystal structure and sufficient evidence in the literature to support mechanism information is given in MACiE. The mechanism steps of the reactions include the function of the catalytic residues involved in the reaction and the mechanism by which substrates are transformed into products.

At the moment there is no database available containing kinetic parameters as quantitative data for individual reaction mechanism steps. Since there is no standard data format for the representation of the mechanism in the literature a structured format is needed to store, easily access and export the data.

For a comprehensive analysis of a biochemical reaction and the enzymatic mechanism detailed information about the reaction mechanism and kinetic parameters for the steps are necessary. Linking of the reaction mechanism data to the overall reaction is essential to relate the information to the corresponding general information of the overall reaction available in SABIO-RK and to other external databases.

REACTION MECHANISM

The data in SABIO-RK are extracted from literature or automatically submitted by wet lab experimenters. The data include information about biochemical reactions and related pathways, reaction participants, enzyme and protein characteristics such as EC number, cellular location, UniProt [10] accession number, molecular weights of protein complexes and subunits and beyond that the description of the protein complex composition (*e.g.* homohexamer described as (subunit)*6). General information about the organism and tissue are linked to the reactions. Additionally, the source of the data is provided *e.g.* by the literature reference. For the biochemical reactions kinetic details are extracted from the literature which includes kinetic parameters (K_m, V_{max}, k_{cat}, K_i values etc.), the type of the kinetic law (Michaelis-Menten, Ping Pong, bi-bi etc.) and the corresponding kinetic law equation. Moreover, the experimental conditions (pH, temperature and buffer) under which the kinetic parameters were determined are available.

Beside the general description and kinetic characterization of the overall biochemical reactions SABIO-RK also collects information about single mechanism steps representing the individual interactions of the single reaction participants (substrates, products, inhibitors, activators, cofactors etc.) with the catalytic proteins or intermediate states of the protein (Fig. 1). Forward and reverse reactions are handled as separate reactions for the representation of the mechanism steps. For example the biochemical reaction "L-serine + L-homocysteine = cystathionine + H_2O" represents an overall reaction in SABIO-RK for which kinetic parameters are stored. Beyond that kinetic parameter are available in the literature for the individual mechanism steps represented as separate reactions (equations 1 – 8) leading to separate mechanism entries in the database linked to the overall reaction mentioned above.

Equations 1 – 8:

(1) E + A → EA
(2) EA → E + A
(3) EA → EX + Q
(4) EX + B → EXB
(5) EXB → EX + B
(6) EXB → EP
(7) EP → E + P
(8) E + P → EP

E, A, B, P, and Q are the representatives of Enzyme, L-serine, L-homocysteine, cystathionine, and H_2O, respectively. X is the intermediate aminoacrylate formed during the catalytic process. Compared to the MACiE database SABIO-RK displays the mechanism steps in an abstract representation, no structural formulae or structural descriptions of catalytic centres of enzymes nor detailed description or representation of chemical interactions between compounds and proteins are indicated.

The SABIO-RK database contains mechanism information both on a qualitative and quantitative level dependent on the information given in the publication. Qualitative information is the representation of the steps to define the order of interactions without any kinetic parameters. This representation is used to understand the types of interactions of all reaction participants with the enzyme and to define different enzyme complexes and intermediate states of the enzyme. If quantitative data are provided, the corresponding kinetic parameters are stored for each single mechanism step. Kinetic parameters represent mainly rate constants but also include participant concentrations and dissociation constants. Rate constants for the forward and reverse reactions like for example $k+1$ and $k-1$ are stored in separate entries because they represent different mechanism steps.

One advantage of the representation of individual reaction mechanism steps is the possibility to define protein-ligand interactions such as the binding of reaction participants to the enzyme or to the enzyme-ligand complexes. These definitions of protein-ligand interactions are helpful in the understanding of the mechanism of inhibitor, activator or cofactor inter-actions.

Each reaction step is related to the overall reaction with its general information (enzyme and protein details, organism, tissue, information source) and is represented equivalently to the overall reaction by its reaction participants (substrates, products, inhibitors, activators etc.), kinetic parameters and corresponding kinetic law information. For each reaction mechanism step individual experimental conditions can be defined independent of the experimental conditions of the overall reaction. For the individual reactions also different kinetic law types and different kinetic law equations compared to the overall reaction can be assigned.

All the mechanism information can be accessed via the overall reaction entries in the SABIO-RK user interface. At the moment web service functions for the mechanism data are not implemented. To search for mechanism entries, the user interface offers a checkbox "Detailed mechanism data (single steps)". Additionally, at least one of the dark blue high-lighted parameters (reactant, pathway, enzyme, organism etc.) on the main search page of the user interface must be selected to search for reaction mechanism details.

On the mechanism details page of the SABIO-RK user interface (Fig. 1) the reaction me-chanism can be displayed by using several predefined automatic layout algorithms [11]. A code using different colours and shapes for the reaction participants helps to easily under-stand the graphical representation of the reactions. The colours used to code for substrates, products, activators, and inhibitors are yellow, blue, green, and red, respectively. Chemical compounds (molecules) are displayed as circles and enzymes and enzyme-compound com-plexes as rectangles. Additionally the graphical representation includes the possibility to select reaction details (triangle) of single graphs representing an individual reaction mechan-ism step to highlight either the reaction equation of the selected step or the kinetic para-meters of the step in a table view. The graphical representation of the mechanism steps was implemented as a Java applet, which provides interactive features to web applications that cannot be provided by HTML alone.

SABIO-RK: Kinetic Data for Reaction Mechanism Steps

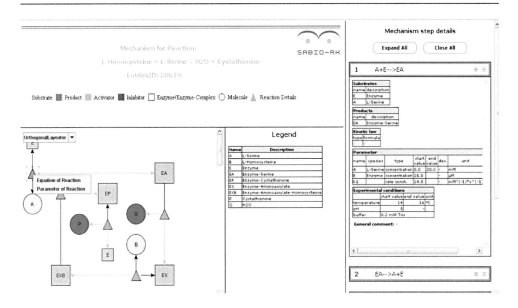

Figure 1. Screenshot of the reaction mechanism details page in the SABIO-RK user interface

Reaction participants of the mechanism steps are explained in more detail in the legend of the mechanism details page. The legend contains a description of the abbreviated step participant names. In the database reaction participants are matched to the reaction participants of the overall reaction. For example a protein complex EAB is related to three different reaction participants of the overall reaction: E, A, and B. They are the representatives of the enzyme, the first substrate (L-serine) and the second substrate (L-homocysteine), respectively, of the reaction mechanism above. This information will be also available as links to the corresponding compound details page in the SABIO-RK user interface in one of the next database releases.

Each single mechanism step is represented in a separate reaction entry containing the detailed information about this step including step equation, step participants details, kinetic parameters and law description, and experimental conditions.

The web-based input interface and the data model of the SABIO-RK database were adapted to the new requirements for the insert and storage of mechanism data. Analogue to the data for main biochemical reactions these mechanism data are manually extracted from literature, are related to overall reaction entries and are manually curated by biological experts. Along the lines of the SABIO-RK data model for overall reactions the mechanism data are represented as similar objects in the database (Fig. 2). Step participants are handled like reactants and modifiers of the main reaction. Mechanism steps are similar to reactions and step parameters to kinetic parameters of overall reactions.

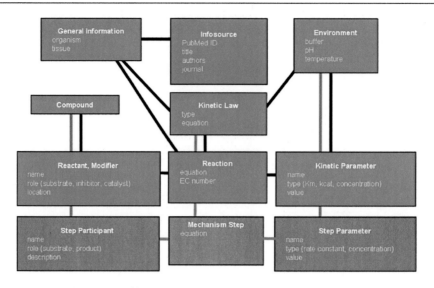

Figure 2. Extended data model (schematic) for representing reaction mechanism

CONCLUSION AND FUTURE PERSPECTIVES

To meet the requirements for a detailed analysis of biochemical reactions and their kinetic behaviour additionally to the kinetic data of the biochemical reactions SABIO-RK now includes a graphical representation of the mechanism of a selected reaction as a survey and also a detailed listing of all individual reaction steps as separate entries. Each reaction step is related to the overall reaction with its general information (enzyme, organism, information source) and is represented equivalently to the overall reaction by its reaction participants (substrates and products), kinetic parameters like for example rate constants and corresponding kinetic law information. For each reaction mechanism step individual experimental conditions can be defined. Beside the graphical representation of the reaction mechanism using several layout algorithms each mechanism step is represented in a separate reaction entry containing the detailed information about this step including step equation, step participants details, kinetic parameters and law description, and corresponding experimental conditions.

The storage and representation of reaction mechanism steps in SABIO-RK offers the possibility to define protein-ligand interactions like for example the binding of reaction participants to the enzyme or to the enzyme-ligand complexes which is important for the representation of signalling reactions in regulatory pathways. The detailed analysis of protein-ligand interactions helps to understand the mechanism of cofactors, inhibitors, or activators in complex networks.

SABIO-RK provides export functions for the overall reactions and their kinetic properties in SBML format [12] to allow the import of the data into simulation and modelling tools. In the future also mechanism steps will be included in the SBML export to combine information and kinetic data from overall reactions with details of reaction mechanism steps.

ACKNOWLEDGEMENT

The authors would like to thank the Klaus Tschira Foundation (KTS) and the German Federal Ministry of Education and Research (BMBF) for financial support.

REFERENCES

[1] Wittig, U., Golebiewski, M., Kania, R., Krebs, O., Mir, S., Weidemann, A., Anstein, S., Saric, J., Rojas, I. (2006) SABIO-RK: integration and curation of reaction kinetics data. *Lecture Notes in Computer Science* **4075**:94 – 103.
doi: http://dx.doi.org/10.1007/11799511_9.

[2] Rojas, I., Golebiewski, M., Kania, R., Krebs, O., Mir, S., Weidemann, A., Wittig, U. (2007) Storing and annotating of kinetic data. *In Silico Biol.* **7**(2 Suppl):S 37 – 44.

[3] Kanehisa, M., Araki, M., Goto, S., Hattori, M., Hirakawa, M., Itoh, M., Katayama, T., Kawashima, S., Okuda, S., Tokimatsu, T., Yamanishi, Y. (2008) KEGG for linking genomes to life and the environment. *Nucleic Acids Res.* **36**:D 480-D 484.

[4] Schomburg, I., Chang, A., Ebeling, C., Gremse, M., Heldt, C., Huhn, G., Schomburg, D. (2004) BRENDA, the enzyme database: updates and major new developments. *Nucleic Acids Res.* **32**:D 431 – 3.
doi: http://dx.doi.org/10.1093/nar/gkh081.

[5] IUBMB: http://www.chem.qmul.ac.uk/iubmb/enzyme/

[6] Fleischmann, A., Darsow, M., Degtyarenko, K., Fleischmann, W., Boyce, S., Axelsen, K.B., Bairoch, A., Schomburg, D., Tipton, K.F., Apweiler, R. (2004) IntEnz, the integrated relational enzyme database. *Nucleic Acids Res.* **32**:D 434-D 437.
doi: http://dx.doi.org/10.1093/nar/gkh119.

[7] Rhea: http://www.ebi.ac.uk/rhea

[8] Vastrik, I., D'Eustachio, P., Schmidt, E., Joshi-Tope, G., Gopinath, G., Croft, D., de Bono, B., Gillespie, M., Jassal, B., Lewis, S., Matthews, L., Wu, G., Birney, E., Stein, L. (2007) Reactome: a knowledge base of biologic pathways and processes. *Genome Biology* **8**:R39.
doi: http://dx.doi.org/10.1186/gb-2007-8-3-r39.

[9]   Holliday, G.L., Almonacid, D.E., Bartlett, G.J., O'Boyle, N.M., Torrance, J.W., Murray-Rust, P., Mitchell, J.B., Thornton, J.M. (2007) MACiE (Mechanism, Annotation and Classification in Enzymes): novel tools for searching catalytic mechanisms. *Nucleic Acids Res.* **35**:D 515-D 520.
        doi: http://dx.doi.org/10.1093/nar/gkl774.

[10]   The UniProt Consortium (2008) The Universal Protein Resource (UniProt). *Nucleic Acids Res.* **36**:D 190-D 195.
        doi: http://dx.doi.org/10.1093/nar/gkm895.

[11]   Kengne Kungne, S. (2008) Konzeption und Implementierung einer webbasierten Visualisierung von Reaktionsmechanismen mit JEE-Technologie, Diploma thesis, University of Applied Sciences Kaiserslautern.

[12]   Hucka, M., Finney, A., Sauro, H.M., Bolouri, H., Doyle, J.C., Kitano, H., Arkin, A.P., Bornstein, B.J., Bray, D., Cornish-Bowden, A., Cuellar, A.A., Dronov, S., Gilles, E.D., Ginkel, M., Gor, V., Goryanin, I.I., Hedley, W.J., Hodgman, T.C., Hofmeyr, J.H., Hunter, P.J., Juty, N.S., Kasberger, J.L., Kremling, A., Kummer, U., Le Novère, N., Loew, L.M., Lucio, D., Mendes, P., Minch, E., Mjolsness, E.D., Nakayama, Y., Nelson, M.R., Nielsen, P.F., Sakurada, T., Schaff, J.C., Shapiro, B.E., Shimizu, T.S., Spence, H.D., Stelling, J., Takahashi, K., Tomita, M., Wagner, J., Wang, J. (2003) The systems biology markup language (SBML): a medium for representation and exchange of biochemical network models. *Bioinformatics* **19**:524 – 31.
        doi: http://dx.doi.org/10.1093/bioinformatics/btg015.

Beilstein-Institut

213

Experimental Standard Conditions of Enzyme Characterizations,
September 13th – 16th, 2009, Rüdesheim/Rhein, Germany

THERMODYNAMIC NETWORK CALCULATIONS APPLIED TO BIOCHEMICAL SUBSTANCES AND REACTIONS

ROBERT N. GOLDBERG

Biochemical Science Division, National Institute of Standards and Technology,
Gaithersburg, Maryland 20899, U.S.A.
and
Department of Chemistry and Biochemistry, University of Maryland,
Baltimore County, Baltimore, MD 21250, U.S.A

E-Mail: robert.goldberg@nist.gov

Received: 16th November 2009 / Published: 14th September 2010

ABSTRACT

Both organic and inorganic chemistry have benefited greatly from the availability of tables of standard enthalpies of formation $\Delta_f H^0$, standard Gibbs energies of formation $\Delta_f G^0$, and standard entropies S^0. These tables of standard thermodynamic properties allow the user to calculate values of enthalpy changes $\Delta_r H^0$, Gibbs energy changes $\Delta_r G^0$, equilibrium constants K, and entropy changes $\Delta_r S^0$ for any reaction in which these standard thermodynamic properties are known for all of the reactants and products. Thus, it is not necessary that actual measurements have been performed on the reaction of interest. While several tables of standard thermodynamic properties have been prepared for biochemical substances, they are not as extensive as the corresponding tables for organic and inorganic substances or as extensive as they might be if all of the available experimental results in the literature had been utilized. Nevertheless, comprehensive tables could be produced by utilizing all of the data {apparent equilibrium constants K' and calorimetrically determined enthalpies of reaction $\Delta_r H(\text{cal})$} in the *Thermodynamics of Enzyme-catalyzed Reactions Database* [1] together with related property values such as standard enthalpies of combustion, entropies and heat capacities, solubilities, enthalpies of solution, pKs, and enthalpies of binding for the substances of interest. This large set of property values can be used to establish a thermodynamic network, i.e., a system of linear equations that can be solved for the desired formation properties. Such an undertaking requires extensive literature work, a substantial amount of analysis and computation on the results of the individual studies, and a careful fitting together of the property values by means of a judicious weighting of the property values. It can be viewed as a very large "jig-saw puzzle" of information. But the proper construction of such a network would serve to bring together a large body of related property values and would be of immense practical value to the scientific community. In addition to the aforementioned utility of the standard thermodynamic properties, there are also the following additional benefits: (1) the consistency or lack of consistency of related results are made visible and needed measurements are identified straightforwardly; (2) the calculated formation properties can be easily updated and revised by a relatively quick calculation so as to allow for the inclusion of both new measurements, remeasurements, corrections, and a revised weighting of results; and (3) the calculated formation properties can, based on structural similarity, serve as the basis for the estimation of property values that have not yet been measured. The table of formation properties and the estimation procedures can also be embedded in computer codes which a user could use to calculate the position of equilibrium of numerous biochemical reactions. The aim of this chapter is to briefly describe the use of these tables, how they are prepared, the current status of existing tables that pertain to biochemical reactions, and to provide a vision of what is possible.

INTRODUCTION

This chapter is an extension of the chapter [2] that I contributed to the *Proceedings of the 3rd International Beilstein Workshop on Experimental Standard Conditions of Enzyme Characterizations (ESCEC)*. An important point made in that chapter was that one can use an equilibrium model together with values of equilibrium constants K and standard enthalpies of reaction $\Delta_r H^0$ for *chemical reactions* (i.e., reactions that involve specific species; both atoms and charges must balance in chemical reactions) to calculate the properties of (*overall*) *biochemical reactions* that involve sums of species. These calculated properties include the apparent equilibrium constant K', the standard transformed enthalpy of reaction $\Delta_r H^0$, the calorimetrically determined enthalpy of reaction $\Delta_r H(\text{cal})$, the changes in binding $\Delta_r N(X)$ for any ligands ($X = H^+$, Mg^{2+}, Ca^{2+}, etc.) associated with the biochemical reaction, the extent of the biochemical reaction ζ', and the concentrations c of all of the species in solution. Thus, one can obtain a *complete thermodynamic picture* of biochemical reactions if one knows the values of the equilibrium constants K and standard enthalpies of reaction $\Delta_r H^0$ for the system of chemical reactions that constitute the biochemical reaction. It is important to appreciate that, unlike the standard thermodynamic properties K and $\Delta_r H^0$, the properties K', $\Delta_r H^0$, $\Delta_r H(\text{cal})$, and $\Delta_r N(X)$ are functions of pH and pX.

One can obtain values of K and $\Delta_r H^0$ directly from the literature, or by means of thermodynamic cycle calculations, or, by using tables of standard thermodynamic properties. Of these three options, tables of standard thermodynamic properties are generally the most useful and convenient. However, these tables do not spontaneously appear in the literature. Indeed, their preparation requires a substantial amount of effort on the part of individuals who have a background both in thermodynamics and in the chemistry of the materials and reactions for which the thermodynamic properties are being evaluated and calculated. The aim of this chapter is to briefly describe the use of these tables, how they are prepared, the current status of existing tables that pertain to biochemical reactions, and to provide a vision of what is possible.

THE BASIS OF THERMODYNAMIC NETWORK CALCULATIONS

Wagman *et al.* [3] have given a detailed description of the data evaluation process used in the preparation of the *NBS Tables of Chemical Thermodynamic Properties*. These tables [3] contain values of the standard enthalpy of formation $\Delta_f H^0$, the standard Gibbs energy of formation $\Delta_f G^0$, the standard entropy S^0, and the standard heat capacity at constant pressure C_P^0 for 92 chemical elements and their compounds. However, the coverage of carbon compounds in these tables stops for substances that contain more than two carbons. Clearly, the scientific principles that underlie the evaluation of the thermodynamic properties of inorganic substances and of small organic molecules are applicable to biochemical substances of any size and complexity. The first and second laws of thermodynamics lead to the definitions of the following state functions: the internal energy U; the enthalpy H; the entropy S; and the Gibbs energy G. A state function has the critically important characteristic that its value depends only on the state of the system (*e. g.*, the temperature T and pressure P) and not upon how the system was brought to that state. The third law of thermodynamics states that the entropy of a pure substance in a perfect crystalline form is zero at $T = 0$. Thus, the laws of thermodynamics provide both the scientific basis for the construction of tables of thermodynamic properties and also give rules for the use of these functions that enables the calculation of the position of equilibrium for chemical reactions. By convention, the values of $\Delta_f H^0$ and $\Delta_f G^0$ are set equal to zero for the thermodynamically stable phase of each element at $T = 298.15$ K, with the exception of phosphorous, for which the white type is selected as the reference form. Another necessary and important convention used in the construction of thermodynamic tables is that $\Delta_f H^0 = \Delta_f G^0 = S^0 = C_P^0 = 0$ for $H^+(aq)$. Additionally, one also needs conventions regarding standard states for pure solids, liquids, gases, and solutions. A summary of the standard states commonly used in chemical thermodynamics is given in Wagman et al [3] table III.

Table 1. Standard thermodynamic properties for several species at $T = 298.15$ K and $P = 1$ bar[a,b]

Substance and state	$\Delta_f H^0$ kJ·mol^{-1}	$\Delta_f G^0$ kJ·mol^{-1}	S^0 J·K^{-1}·mol^{-1}
$O_2(g)$	0	0	205.152
$H^+(aq)$	0	0	0
$H_2(g)$	0	0	130.680
$OH^-(aq)$	-230.015	-157.220	-10.90
$N_2(g)$	0	0	191.609
$H_2O(l)$	-285.830	-237.14	69.95
P(cr, white)	0	0	41.09
$HPO_4^{2-}(aq)$	-1299.0	-1096.0	-33.5
C(cr)	0	0	5.74
$CO_2(g)$	-393.51	-394.37	213.785
$Mg^{2+}(aq)$	-467.0	-455.4	-137.0
$AMP^{2-}(aq)$, $C_{10}H_{12}N_5O_7P^{2-}$	-1635.37	-1040.45	214.5
$HAMP^-(aq)$, $C_{10}H_{13}N_5O_7P^-$	-1629.97	-1078.86	361.4
$H_2AMP(aq)$, $C_{10}H_{14}N_5O_7P$	-1648.07	-1101.63	377.1
$MgAMP(aq)$, $C_{10}H_{12}N_5O_7PMg$	-2091.07	-1511.77	168.8
$ADP^{3-}(aq)$, $C_{10}H_{12}N_5O_{10}P^-$	-2626.54	-1906.13	207.4
$HADP^{2-}(aq)$, $C_{10}H_{13}N_5O_{10}P^-$	-2620.94	-1947.10	363.6
$H_2ADP^-(aq)$, $C_{10}H_{14}N_5O_{10}P^-$	-2638.54	-1971.98	388.0
$MgADP^-(aq)$, $C_{10}H_{12}N_5O_{10}P_2Mg^-$	-3074.54	-2388.06	223.1
$MgHADP(aq)$, $C_{10}H_{13}N_5O_{10}P_2Mg$	-3075.44	-2416.77	316.3
$ATP^{4-}(aq)$, $C_{10}H_{12}N_5O_{13}P^-$	-3619.21	-2768.10	182.8
$HATP^{3-}(aq)$, $C_{10}H_{13}N_5O_{13}P^-$	-3612.91	-2811.48	349.4
$H_2ATP^{2-}(aq)$, $C_{10}H_{14}N_5O_{13}P^-$	-3627.91	-2838.18	388.7
$MgATP^{2-}(aq)$, $C_{10}H_{12}N_5O_{13}P_3Mg^{2-}$	-4063.31	-3258.77	240.9
$MgHATP^-(aq)$, $C_{10}H_{13}N_5O_{13}P^-Mg^-$	-4063.01	-3287.59	338.5
$Mg_2ATP(aq)$, $C_{10}H_{12}N_5O_{13}P_3Mg_2$	-4519.51	-3729.52	191.6

[a]The property values for the AMP, ADP, and ATP species are from Boerio-Goates et al. [4]. The remaining property values are from the CODATA Tables [5]. The standard state for aqueous solutes is the hypothetical ideal solution of unit molality.
[b]The formation reaction for a neutral substance such as aqueous glucose ($C_6H_{12}O_6$) is: $C(cr) + 6 H_2(g) + 3 O_2(g) = C_6H_{12}O_6(cr)$. The formation reaction for a charged species such as $AMP^{2-}(aq)$ is: $10 C(cr) + 7 H_2(g) + 2.5 N_2(g) + 3.5 O_2(g) + P(cr) = C_{10}H_{12}N_5O_7$. $P^{2-}(aq) + 2 H^+(aq)$.

A sample table of thermodynamic properties is shown in Table 1. One can use the values in tables of this type to calculate standard Gibbs energy changes $\Delta_r G^0$, standard enthalpy changes $\Delta_r H^0$, and standard entropy changes $\Delta_r S^0$ for chemical reactions, e.g.,

$$\nu_A A + \nu_B B = \nu_C C + \nu_D D. \tag{1}$$

Here ν_A, ν_B, ν_C, and ν_D are the stoichiometric numbers corresponding to the respective species A, B, C, and D. The equations needed to calculate $\Delta_r G^0$, $\Delta_r H^0$, and $\Delta_r S^0$ from the tabulated values of $\Delta_f G^0$, $\Delta_f H^0$, and S^0 are

$$\Delta_r G^0 = \nu_C \Delta_f G_C^0 + \nu_D \Delta_f G_D^0 - \nu_A \Delta_f G_A^0 - \nu_B \Delta_f G_B^0, \tag{2}$$

$$\Delta_r H^0 = \nu_C \Delta_f H_C^0 + \nu_D \Delta_f H_D^0 - \nu_A \Delta_f H_A^0 - \nu_B \Delta_f H_B^0, \tag{3}$$

$$\Delta_r S^0 = \nu_C S_C^0 + \nu_D S_D^0 - \nu_A S_A^0 - \nu_B S_B^0. \tag{4}$$

Here, the subscripts for $\Delta_f G^0$, $\Delta_f H^0$, and S^0 refer to the species A, B, C, and D. The equilibrium constant K for a reaction can be calculated from $\Delta_r G^0$ by using the well-known relation

$$K = \exp\left(-\frac{\Delta_R G^0}{RT}\right) \tag{5}$$

Additionally one has the relation

$$\Delta_f G^0 = \Delta_f H^0 - T\Delta_f S^0. \tag{6}$$

Note that if one knows the value of $\Delta_r H^0$ for a given reaction and the values of $\Delta_f H^0$ for all of the species in that reaction with the exception of one of them, eq. (3) can be used to calculate the value of $\Delta_f H^0$ for that one species. The same principle applies to the use of eq. (2) for the calculation of $\Delta_f G^0$ and to the use of eq. (4) to calculate S^0 for a given species. Thus, eqs. (2) to (6) can be used to calculate values of $\Delta_f G^0$, $\Delta_f H^0$, and S^0 from experimentally determined values of K, $\Delta_r H^0$, and $\Delta_r S^0$. Specific categories of data that comprise the aforementioned three properties are: standard enthalpies of combustion $\Delta_c H^0$, standard enthalpies of solution $\Delta_{sol} H^0$, saturation molalities $m(\text{sat})$ or solubilities s, activity coefficients γ, equilibrium constants K, enthalpies of reaction $\Delta_r H^0$, standard entropies S^0, standard heat capacities C_P^0 standard heat capacity changes $\Delta_r C_P^0$, and standard electrode potentials E^0. One can calculate $\Delta_r G^0$ from E^0 by means of the equation

$$\Delta_r G^0 = -|\nu_e| F E^0, \tag{7}$$

where $|\nu_e|$ is the number of electrons in the half reaction for the electrochemical cell and F is the Faraday constant. Apparent equilibrium constants K' and calorimetrically determined enthalpies of reaction $\Delta_r H(\text{cal})$ can also be used in these calculations. However, it is first necessary to calculate values of K and $\Delta_r H^0$ for a chemical reference reaction from the measured values of K' and $\Delta_r H(\text{cal})$. It should also be noted that equilibrium constants can be determined by many different methods – chromatography, electrochemistry, chemical analysis, and a wide variety of spectroscopic methods. Equilibrium constants can also be obtained from kinetic experiments by means of Haldane relations. $\Delta_r H^0$ can be obtained both by means of calorimetry and by measuring K as a function of temperature:

$$\Delta_r H^0 = RT^2(\partial \ln K / \partial T)_P. \tag{8}$$

The most important points follow. Firstly, a wide variety of chemical thermodynamic data must be evaluated in order to construct tables of standard thermodynamic properties. Secondly, this wide variety of data can be reduced to just a few properties (omitting C_P^0 and $\Delta_r C_P^0$), namely $\Delta_r G^0$, $\Delta_r H^0$, and S^0. The collection of such evaluated data is generally referred to as a *reaction catalogue*. Thirdly, since $\Delta_r G^0$, $\Delta_r H^0$, and S^0 are related via eqs. (2), (3), (4), and (6), values of $\Delta_f H^0$, $\Delta_r G^0$, and S^0 can be calculated by solving the system of linear equations that comprise the reaction catalogue. These calculations are often referred to as *thermodynamic network calculations*. Lastly, these standard thermodynamic properties can be used to calculate property values (*e. g.*, K, $\Delta_r G^0$, $\Delta_r H^0$) for many reactions, including those that have not been studied directly.

THERMOCHEMISTRY OF ADENOSINE AND THE ATP SERIES

A useful and interesting example of the calculation of standard thermodynamic properties is now considered. In 1992, Alberty and Goldberg [6] calculated the standard thermodynamic properties of the ATP series which included the H^+ and Mg^{2+} species that constitute the reactants AMP, ADP, and ATP as well as their hydrolysis reactions. In the absence of formation property data for adenosine(aq), they [6] adopted the convention that $\Delta_f H^0 = \Delta_f G^0 = 0$ for adenosine(aq). However, the need for this convention was eliminated when Boerio-Goates *et al.* [4] reported values for the standard enthalpy of combustion of adenosine(cr), the saturation molality of adenosine(cr), and heat capacities for adenosine(cr) over the temperature range 11 K to 328 K.

Table 2. Thermodynamic quantities at $T = 298.15$ K and pressure $P = 1$ bar for reactions and for substances used in thermodynamic network calculations. The empirical formulas of the various substances are: adenine, $C_5H_5N_5$; adenosine, $C_{10}H_{13}N_5O_4$; and D-ribose, $C_5H_{10}O_5$. The standard state for aqueous solutes is the hypothetical ideal solution of unit molality

Entry No.	Substance or Reaction	$\Delta_r H^0$	$\Delta_r G^0$	S^0
		kJ·mol^{-1}	kJ·mol^{-1}	J·K^{-1}·mol^{-1}
1	$C_{10}H_{13}N_5O_4(cr) + 11.25\ O_2(g) = 10\ CO_2(g) + 6.5\ H_2O(l) + 2.5\ N_2(g)$	-5139.4 ± 3.3^a		
2	$C_{10}H_{13}N_5O_4(cr)$			289.57 ± 0.6^b
3	$C_{10}H_{13}N_5O_4(cr) = C_{10}H_{13}N_5O_4(aq)$	32.26 ± 0.17^c	9.89 ± 0.02^d	
4	$C_5H_5N_5(cr) + 6.25\ O_2(g) = 5\ CO_2(g) + 2.5\ H_2O(l) + 2.5\ N_2(g)$	-2779.0 ± 1.3^e		
5	$C_5H_5N_5(cr)$			151.0 ± 3^f
6	$C_5H_5N_5(cr) = C_5H_5N_5(aq)$	33.47 ± 1.0^g	11.98 ± 0.3^h	
7	$C_{10}H_{13}N_5O_4(aq) + H_2O(l) = C_5H_5N_5(aq) + C_5H_{10}O_5(aq)$		-9.84 ± 0.4^i	
8	$C_5H_{10}O_5(cr) + O_2(g) = 5\ CO_2(g) + 5\ H_2O(aq)$	-2349.5 ± 1.0^j		
9	$C_5H_{10}O_5(cr) = C_5H_{10}O_5(aq)$	13.2 ± 0.3^k	-8.7 ± 0.5^l	
10	$C_5H_{10}O_5(cr)$			175.7 ± 0.6^m

[a]Based on the standard enthalpy of combustion $\Delta_c H^0$ of adenosine(cr) reported by Boerio-Goates *et al.* [4].
[b]Based on the standard entropy S^0 of adenosine(cr) reported by Boerio-Goates *et al.* [4].
[c]Based on the standard enthalpy of solution $\Delta_{sol}H$ of adenosine(cr) reported by Stern and Oliver [7].
[d]Based on the saturation molality m_{sat} of adenosine(cr) reported by Boerio-Goates *et al.* [4].
[e]Based on the standard enthalpy of combustion $\Delta_c H^0$ of adenine(cr) reported by Kirklin and Domalski [8].
[f]Based on the standard entropy S^0 of adenine(cr) reported by Stiehler and Huffman [9].
[g]Based on the standard enthalpy of solution $\Delta_{sol}H^0$ of adenine(cr) reported by Kilday [10].
[h]Based on the mean of the values of $m(sat)$ for adenine(cr) in water obtained from four investigations [11–14].
[i]Based on the value of K' measured by Camici *et al.* [15] for the reaction: {adenosine(aq) + H$_2$O(l) = adenine(aq) + D-ribose(aq)}.
[j]Based on the standard enthalpy of combustion $\Delta_c H^0$ of ribose(cr) reported by Colbert *et al.* [16].
[k]Based on the standard enthalpy of solution $\Delta_{sol}H^0$ of ribose(cr) reported by Jasra and Ahluwalia [17] and by Stern and Hubler [18].
[l]Based on the saturation molality m_{sat} of ribose(cr) reported by Goldberg and Tewari [19].
[m]Based on the third law entropy S^0 of ribose(cr) reported by Boerio-Goates *et al.* [4].

The heat capacities led, by application of the third law of thermodynamics, to the standard entropy S^0 for adenosine(cr) at $T = 298.15$ K. The data that enter into these calculations are given as the first three entries in Table 2. Entry no. 1 contains the data for the enthalpy of combustion $\Delta_c H^0$ of adenosine(cr). Entry No. 2 contains the data for the standard entropy S^0 of adenosine(cr). And entry no. 3 contains the data for both the enthalpy of solution $\Delta_{sol}H^0$ and the standard Gibbs energy of solution $\Delta_{sol}G^0$ of adenosine(cr). The latter quantity was calculated by using the formula

$$\Delta_{sol}G^0 = -RT\ln\{\gamma \cdot m(sat) \cdot a_w^N\}. \tag{9}$$

Here $m(sat)$ is the saturation molality {i.e., the amount of adenosine that is soluble in 1 kg of H$_2$O(l)} of adenosine(cr) in H$_2$O(l), γ is the activity coefficient of adenosine(aq), a_w is the activity of water in the saturated solution, and N is the number of waters of hydration associated with adenosine(cr). Since adenosine(cr) is anhydrous, N is equal to zero, and $a_w^N = 1$. The steps in the calculation of $\Delta_f H^0$, $\Delta_f G^0$, and S^0 for adenosine(aq) are now shown schematically:

$\Delta_c H^0$ of adenosine(cr) + values of $\Delta_f H^0$ for $H_2O(l)$ and $CO_2(g)$ → $\Delta_f H^0$ for adenosine(cr),

S^0 of adenosine(cr) + values of S^0 for C(cr), $H_2(g)$, $N_2(g)$, and $O_2(g)$ → $\Delta_f S^0$ for adenosine(cr),

$\Delta_f H^0 + \Delta_f S^0$ for adenosine(cr) → $\Delta_f G^0$ for adenosine(cr),

$\Delta_f H^0$ for adenosine(cr) + $\Delta_{sol} H^0$ for adenosine(cr) → $\Delta_f H^0$ for adenosine(aq),

$\Delta_f G^0$ for adenosine(cr) + $\Delta_{sol} G^0$ for adenosine(cr) → $\Delta_f G^0$ for adenosine(aq),

$\Delta_f H^0$ and $\Delta_f G^0$ for adenosine(aq) → $\Delta_f S^0$ for adenosine(aq),

$\Delta_f S^0$ for adenosine(aq) + values of S^0 for C(cr), $H_2(g)$, $N_2(g)$, and $O_2(g)$ → S^0 for adenosine(aq).

Eqs. (2), (3), (4), and (6) were the only equations used in the above schematic calculation. However, in addition to the property values shown in entries no. 1 to 3, accurate property values for key substances such as $H_2O(l)$, $CO_2(g)$, C(cr), $N_2(g)$, and $O_2(g)$ were also required. These values were taken from the CODATA Tables [5]. By using this pathway, the calculated formation properties of adenosine(aq) are:

$\Delta_f H^0 = -(621.3 \pm 3.6)$ kJ·mol^{-1}, $\Delta_f G^0 = -(194.5 \pm 3.6)$ kJ·mol^{-1}, and
$S^0 = (364.6 \pm 0.7)$ J·K^{-1}·mol^{-1}.

A second thermodynamic pathway that leads to $\Delta_f G^0$ for adenosine(aq) uses entries no. 4 to 10 in Table 2. The basis for these entries is given in the footnotes in that table. The steps in the calculation of $\Delta_f G^0$ for adenosine(aq) are similar to those shown above except that they are based on the formation properties of adenine and of ribose. A link from these two substances to adenosine is obtained by means of the value of K' measured by Camici *et al.* [15] for the reaction:

$$\text{adenosine(aq)} + H_2O(l) = \text{adenine(aq)} + \text{D-ribose(aq)}. \quad (10)$$

The value $\Delta_f G^0 = -(192.4 \pm 6.3)$ kJ·mol^{-1} for adenosine(aq) obtained by this second pathway, is in excellent agreement with the result obtained by use of the first pathway described above. However, since the situation is not always this favourable, one must, in general, either select one set of data over another over another or one must weight the data in some reasonable way in order to obtain a set of "best" property values.

It should be noted that the uncertainties in the above values of $\Delta_f G^0$ and of $\Delta_f H^0$ can be attributed largely to the uncertainties in the values of the standard enthalpies of combustion of adenosine(cr), adenine(cr), and D-ribose(cr). Thus, one expects to see uncertainties in the values of $\Delta_f G^0$ and of $\Delta_f H^0$ on the order of several kJ·mol^{-1} for many aqueous species. However, in many cases, much of the uncertainty will cancel when the tabulated values of $\Delta_f G^0$ and of $\Delta_f H^0$ are used to calculate the desired values of $\Delta_r G^0$ and of $\Delta_r H^0$ for reactions that are occurring in aqueous media. This cancellation occurs when all of the property values used in the calculation are independent of standard enthalpies of combustion and when the uncertainties in the measured property values are not too large. In any case, the tabulated values of standard formation properties should largely reproduce, within their estimated uncertainties and weights, the property values that were measured. Clearly, one needs reliable standard enthalpies of combustion as a basis for the values of standard formation properties of biochemical species in aqueous media. However, once a sufficient number of such values have been established, the measured values of K' and $\Delta_r H(\text{cal})$ (see: http://xpdb.nist.gov/enzyme_thermodynamics/) provide numerous thermodynamic pathways that lead to the standard thermodynamic properties of many biochemical substances. The preceding discussion emphasizes the importance of uncertainties in thermodynamic network calculations. It is unfortunate that many thermodynamic investigations give little information on uncertainties other than the estimated uncertainty due to random error, i.e., the standard deviation associated with a measurement.

Table 3. Thermodynamic quantities selected by Alberty and Goldberg [6] for aqueous AMP, ADP, ATP, and phosphate reactions at $T = 298.15$ K. $P = 1$ bar, and ionic strength $I = 0$

Reaction	$\Delta_r HE$	$\Delta_r GE$
	kJ·mol^{-1}	kJ·mol^{-1}
$HAMP^- = H^+ + AMP^{2-}$	-5.4	38.42
$H_2AMP = H^+ + HAMP^-$	18.1	22.78
$MgAMP = Mg^{2+} + AMP^{2-}$	-11.3	15.93
$HADP^{2-} = H^+ + ADP^{3-}$	-5.6	40.98
$H_2ADP^- = H^+ + HADP^{2-}$	17.6	24.89
$MgADP^- = Mg^{2+} + ADP^{3-}$	-19.0	26.54
$MgHADP = Mg^{2+} + HADP^{2-}$	-12.5	14.27
$HATP^{3-} = H^+ + ATP^{4-}$	-6.3	43.38
$H_2ATP^{2-} = H^+ + HATP^{3-}$	15.0	26.71
$MgATP^{2-} = Mg^{2+} + ATP^{4-}$	-22.9	35.28
$MgHATP^- = Mg^{2+} + HATP^{3-}$	-16.9	20.72
$Mg_2ATP = Mg^{2+} + MgATP^{2-}$	-10.8	15.35
$ATP^{4-} + H_2O(l) = ADP^{3-} + HPO_4^{2-} + H^+$	-20.5	3.03
$ADP^{3-} + H_2O(l) = AMP^{2-} + HPO_4^{2-} + H^+$	-22.0	6.73
$AMP^{2-} + H_2O(l) = \text{adenosine} + HPO_4^{2-}$	0.9	-13.00

By having reliable values of $\Delta_f H^0$, $\Delta_f G^0$, and S^0 for adenosine(aq), it is then possible to use the property values selected by Alberty and Goldberg [6] (see table 3) for aqueous AMP, ADP, and ATP reactions in order to calculate the property values for the AMP, ADP, and ATP species shown in Table 1. Again, in performing these calculations, it was necessary to use auxiliary property data for Mg^{2+}(aq) and for HPO_4^{2-}(aq). Values for these properties were also taken from the CODATA Tables [5]. Note that the calculation of the standard thermodynamic properties of adenosine(aq) required the use of several different properties for adenosine(cr): $\Delta_c H^0$, S^0, $\Delta_{sol} H^0$, and $\Delta_{sol} G^0$. The value of $\Delta_{sol} G^0$ required the use of eq. (8) and a knowledge of several additional properties, namely, m(sat), γ, and N, the number of waters of hydration associated with adenosine(cr). However, once the standard thermodynamic properties of adenosine(aq) were established, the calculation of the standard formation properties for the AMP, ADP, and ATP species required only the values of $\Delta_r H^0$ and $\Delta_r G^0$ for reactions occurring in aqueous solution. Thus, by having obtained a thermodynamic connection of one substance, namely adenosine(aq), to the elements, one was then able to establish the formation properties of many other biochemical substances. Similar tactics are applicable to many other biochemical substances for which values of $\Delta_c H^0$, S^0, $\Delta_{sol} H^0$, and $\Delta_{sol} G^0$ either have been or can be measured.

Discussion

Clearly, these calculations are very detailed and great care is required to make certain that property values have been properly extracted from the literature and treated properly. This often involves the recalculation of the results reported in the literature. However, with modern computers, the solution of thermodynamic networks is made substantially easier and is accomplished much more rapidly than the old, sequential method which was used above for illustrative purposes. The essential point is that once the thermodynamic catalogue has been established (Tables 2 and 3, together with the necessary auxiliary property values, constitute a catalogue); the calculation of values of $\Delta_f H^0$, $\Delta_f G^0$, and S^0 can be done almost instantaneously. Thus, when new and better experimental results appear, they can be easily incorporated into the catalogue and new values of $\Delta_f H^0$, $\Delta_f G^0$, and S^0 can be obtained with little effort.

Thus, the central problem becomes the construction of the master thermodynamic catalogue for biochemical species and reactions. The evaluation or weighting of the data is clearly an important part of any effort of this type.

Based on the above discussion, it is highly desirable that the master catalogue contain values of $\Delta_r H^0$, $\Delta_r G^0$, and S^0 for as many biochemical substances as possible. A good start for a list of biochemical substances to be included in the master catalogue is the ~1200 biochemical reactants found in the *Thermodynamics of Enzyme-catalyzed Reactions Database* [1]. As seen above, the thermodynamic properties that are of interest are: standard enthalpies of combustion $\Delta_c H^0$, standard enthalpies of solution $\Delta_{sol} H^0$, saturation molalities m(sat) or solubilities s, activity coefficients γ, equilibrium constants K, enthalpies of reaction $\Delta_r H^0$, and standard entropies S^0. In some cases, standard electrode potentials E^0 may prove valuable in the calculation of values of $\Delta_r G^0$. While of interest, but of secondary importance to values of $\Delta_r G^0$, $\Delta_r H^0$, and S^0, are standard heat capacities C_P^0 and standard heat capacity changes $\Delta_r C_P^0$ for reactions. An examination of the *Thermodynamics of Enzyme-catalyzed Reactions Database* [1] shows that many of the most useful pathways will involve the use of apparent equilibrium constants K' and calorimetrically determined enthalpy changes $\Delta_r H$(cal) for enzyme-catalyzed reactions. As stated earlier, values of K and $\Delta_r H^0$ can be calculated from the experimentally reported values of K' and $\Delta_r H$(cal). This can be done by means of chemical equilibrium modelling calculations [20]. However, in performing such calculations, in general, one requires the knowledge of the pKs and corresponding $\Delta_r H^0$ values for the H^+ and Mg^{2+} and, possibly, Ca^{2+} binding constants for the reactions under consideration. Clearly, the amount of literature work required both to extract and to *evaluate* all of this data is very considerable and will require several years of effort by skilled workers. But, assuming that it has been done, one is still left with the task of performing the network calculations, including the assignment of weights, in order to obtain a set of "best" standard thermodynamic properties. In two previous evaluation efforts [19, 21] involving biochemical reactions, the computer codes described by Pedley *et al.* [22, 23] and by Garvin *et al.* [24] were used to solve the thermodynamic networks and to obtain the desired formation properties.

The above discussion has outlined both the importance of creating the master thermodynamic catalogue for biochemical substances and reactions as well as the very formidable logistical matter of collecting and evaluating the necessary data and of performing the essential calculations. However, once the master catalogue has been established and a reasonable set of weights assigned to the entries, the calculation of the desired standard thermodynamic properties can be accomplished almost instantaneously. Thus, as new data is published in the literature or if existing entries in the master catalogue need to be corrected or weighted differently, the calculation of the revised standard thermodynamic properties is easily done.

Finally, once one has tables of standard thermodynamic properties in hand, one clearly wants to have them in a database that can be accessed by application programs. One application that many users should find particularly useful is to be able to simply write a biochemical reaction and specify the conditions of the reaction, namely, T, I, pH, and pMg. The program then obtains the required data from a table of standard thermodynamic properties and returns all of the desired information that the user might wish to know, namely, K', $\Delta_r G'^0$, $\Delta_r H'^0$, $\Delta_r H$(cal), and $\Delta_r N(X)$ for any ligand ($X = H^+$, Mg^{2+}, Ca^{2+}, etc.) associated with the biochemical reaction under a specified set of conditions. Values of the standard electrode potential E and the apparent standard electrode potential E'^0 can also be calculated.

A *Mathematica* package that does much of this is *Basic Data for Biochemistry* [25, 26]. This package is available for download from: http://library.wolfram.com/infocenter/MathSource/ 797. Additionally, the *Mathematica* package *BioEqCalc* [20] is useful in calculating the concentrations of species and reactants once one has entered the reactions and the values of the equilibrium constants for the reactions. *BioEqCalc* can also be used to calculate values of K', $\Delta_r G'^0$, $\Delta_r H'^0$, $\Delta_r H(\text{cal})$, $\Delta_r N(X)$ as well as calculate the values of K and $\Delta_r H^0$ for chemical reference reactions from values of K', $\Delta_r H'^0$, and $\Delta_r H(\text{cal})$.

Table 4. Published tables of standard thermodynamic properties for biochemical substances[a]

Publication	No. of substances/ species	Content	Properties in table
Krebs *et al.* [27]	130	General biochemistry	$\Delta_f G^0$
Wilhoit [28]	246	General biochemistry	$\Delta_f G^0$, $\Delta_f H^0$, S^0, C_P^0
Thauer *et al.* [29][b]	165	General biochemistry	$\Delta_f G^0$
Goldberg and Tewari [19]	74	Carbohydrates (pentoses and hexoses) and their monophosphates	$\Delta_f G^0$, $\Delta_f H^0$, S^0, C_P^0
Miller and Smith-Magowan [21]	26	Krebs cycle compounds	$\Delta_f G^0$
Ould-Moulaye *et al.* [30, 31]	109	Purines, pyrimidines, nucleosides, nucleotides, and nucleotide phosphates	$\Delta_f G^0$, $\Delta_f H^0$, S^0, C_P^0
Alberty [25, 26]	199	General biochemistry – aqueous species	$\Delta_f G^0$, $\Delta_f H^0$

[a]The user of any of these tables is cautioned about the risks in combining data from two or more different tables to calculate property values for a given reaction. Specifically, while each of the aforementioned thermochemical tables is presumably internally consistent, the values may not be consistent with the values given in another table. Thus, serious errors can result if values from two or more tables are combined to calculate property values for a given reaction. Two of the above tables [21, 29] do not appear to have taken into account the difference between biochemical reactants and chemical species.
[b]Many of the values of $\Delta_f G^0$ were taken from other tables.

Table 4 summarizes the publications that contain substantial numbers of entries for the standard thermodynamic properties of biochemical substances. The most extensive of the thermochemical tables (Wilhoit [28]) contains data for 246 species. However, given the fact that there are ~1200 biochemical reactants in the *Thermodynamics of Enzyme-catalyzed Reactions Database* [1], it is clear that a considerable amount of effort is still needed in order to extract what is a very substantial amount of useful data from the existing literature. But, if this task were to be completed, one would have available not only a very useful table of standard thermodynamic properties, but also, as pointed out above, a reaction catalogue that could be updated relatively easily as new and improved data became available. The reaction catalogue would also serve as a guide as to which measurements were most valuable both in terms of resolving discrepancies in already published results and in where the measurement needs were most urgent for new biochemical reactions and reactants. The tables of standard formation properties would also allow one to apply the Benson group-contribution method [32 – 36] in order to obtain characteristic values for distinct chemical groups. These group values can then be used to estimate the properties of substances that have not been the subject of a direct study. The Benson method is particularly useful for the estimation of values of $\Delta_r S^0$. These estimated values of $\Delta_r S^0$, can then be used with measured values of $\Delta_r H^0$ to calculate values of $\Delta_r G^0$ and K for reactions where the measurement of the equilibrium constant is either difficult or essentially impossible due to the fact that the reaction of interest goes largely to completion. An obvious, but important point is that the usefulness of tables of standard thermodynamic properties substantially increases with the number of substances in the tables. Given the increased interest in quantitative biochemistry (*e.g.*, metabolic engineering, metabolic control analysis, and systems biology) and the direct tie to kinetic data via the Haldane relations, the important task of creating such a reaction catalogue and the accompanying tables of standard formation properties is obviously one that needs to be undertaken and accomplished. Clearly, this task will require a concerted effort by several individuals or groups over an extended period of time. But, given the size and never-ending nature of the task, it would be desirable to have a central depository where contributions to the master thermodynamic catalogue can be deposited. An important part of such a central depository would be to insure a very high, but still reasonable degree of quality control over any input contributed to it.

ACKNOWLEDGMENT

I thank Professor Robert A. Alberty for his helpful comments.

GLOSSARY OF SYMBOLS AND TERMINOLOGY[A]

Symbol	Name	Units
c	concentration	$mol \cdot dm^{-3}$
C_P^0	standard heat capacity at constant pressure	$J \cdot K^{-1} \cdot mol^{-1}$
$\Delta_r C_P^0$	standard heat capacity change for a reaction at constant pressure	$J \cdot K^{-1} \cdot mol^{-1}$
E^0	standard electrode potential of a cell	V
E'^0	standard apparent electrode potential of a cell	V
F	Faraday constant (96485.3399 C mol-1)	$C \cdot mol^{-1}$
G	Gibbs energy	J
$\Delta_r G^0$	standard Gibbs energy change for a reaction	$J \cdot mol^{-1}$
$\Delta_r G'^0$	standard transformed Gibbs energy change for a biochemical reaction	$J \cdot mol^{-1}$
H	enthalpy	J
$\Delta_r H^0$	standard enthalpy change for a reaction	$J \cdot mol^{-1}$
$\Delta_r H'^0$	standard transformed enthalpy change for a biochemical reaction	$J \cdot mol^{-1}$

Symbol	Name	Units
$\Delta_r H(cal)$	calorimetrically determined enthalpy change for a biochemical reaction	$J \cdot mol^{-1}$
I	ionic strength	$mol \cdot dm^{-3}$
K	equilibrium constant	dimensionless
K'	apparent equilibrium constant	dimensionless
$m(sat)$	saturation molality	$mol \cdot kg^{-1}$
N	number of waters of hydration associated with a crystalline substance	dimensionless
$\Delta_r N(X)$	change in binding of species X for a biochemical reaction	dimensionless
P	pressure	bar
pH	$-\log_{10}[H^+]$	dimensionless
pK	$-\log_{10}K$	dimensionless
pMg	$-\log_{10}[Mg^{2+}]$	dimensionless
R	gas constant ($8.314472\ J \cdot K^{-1} \cdot mol^{-1}$)	$J \cdot K^{-1} \cdot mol^{-1}$
s	solubility	$mol \cdot dm^{-3}$
S	entropy	$J \cdot K^{-1}$
S^0	standard entropy	$J \cdot K^{-1} \cdot mol^{-1}$
$\Delta_r S^0$	standard entropy change for a reaction	$J \cdot K^{-1} \cdot mol^{-1}$
T	temperature	K
U	internal energy	J
γ	activity coefficient	dimensionless
ν	stoichiometric number	dimensionless
$|\nu_e|$	number of electrons in a half reaction	dimensionless
ξ'	extent of a biochemical reaction	mol

[a]The subscript "r" following a "Δ" denotes a reaction. Subscripts "c", "f", and "sol" are used, respectively, to denote combustion, formation, and solution reactions. A subscript "m" is often used to denote molar quantities, but this is not necessary if the units or context are clearly specified. The symbols (aq), (cr), and (l) denote, respectively, an aqueous species, a crystalline substance, and a liquid substance. An "M" is often used as an abbreviation for $mol \cdot dm^{-3}$ or $mol \cdot L^{-1}$.

REFERENCES

[1]   Goldberg, R.N., Tewari, Y.B., Bhat, T.N. (2004) Thermodynamics of Enzyme-Catalyzed Reactions – A Database for Quantitative Biochemistry. *Bioinformatics* **20**:2874 – 2877.
        doi: http://dx.doi.org/10.1093/bioinformatics/bth314
        http://xpdb.nist.gov/enzyme_thermodynamics/.

[2]   Goldberg, R.N. (2008) Thermodynamic property values for enzyme-catalyzed reactions. In: *Proceedings of the 3rd International Beilstein Workshop on Experimental Standard Conditions of Enzyme Characterizations* (Hicks, M.G., Kettner, C., Eds). Logos Verlag Berlin, Berlin; pp. 47 – 62.

[3]   Wagman, D.D., Evans, W.H., Parker, V.B., Schumm, R.H., Halow, I., Bailey, S.M., Churney, K.L., Nuttall, R.L. (1982) The NBS tables of chemical thermodynamic properties. *J. Phys. Chem. Ref. Data* **11**, Supplement 2.

[4]   Boerio-Goates, J., Francis, M.R., Goldberg, R.N., Ribiero da Silva, M.A.V., Ribiero da Silva, M.D.M.C., Tewari, Y.B. (2001) Thermochemistry of adenosine. *J. Chem. Thermodyn.* **33**:929 – 947.
        doi: http://dx.doi.org/10.1006/jcht.2001.0820.

[5]   Cox, J. D., Wagman, D.D., Medvedev, V.A. (1989) *CODATA Key Values for Thermodynamics*. Hemisphere, New York.

[6]   Alberty, R.A., Goldberg, R.N. (1992) Standard thermodynamic formation properties for the adenosine triphosphate series. *Biochemistry* **31**:10610 – 10615.
        doi: http://dx.doi.org/10.1021/bi00158a025.

[7]   Stern, J.H., Oliver, D.R. (1980) Thermodynamics of Nucleoside-Solvent Interactions: Inosine and Adenosine in Water and in 1 m Ethanol between 25 and 35 $^{\circ}$C. *J. Chem. Eng. Data* **25**:221 – 223.
        doi: http://dx.doi.org/10.1021/je60086a033.

[8]   Kirklin, D.R., Domalski, E.S. (1983) Enthalpy of combustion of adenine. *J. Chem. Thermodyn.* **15**: 941 – 947.
        doi: http://dx.doi.org/10.1016/0021-9614(83)90127-1.

[9]   Stiehler, R.D., Huffman, H.M. (1935) Thermal Data. V. The Heat Capacities, Entropies and Free Energies of Adenine, Hypoxanthine, Guanine, Xanthine, Uric Acid, Allantoin and Alloxan. *J. Am. Chem. Soc.* **57**:1741 – 1743.
        doi: http://dx.doi.org/10.1021/ja01312a072

[10]   Kilday, M.V. (1978) Enthalpies of solution of nucleic-acid bases. 1. Adenine in water. *J. Res. Natl. Bur. Stand. (U. S.)* **83**:347 – 370.

[11]   Stockx, J., Van Aert, A., Clauwaert, J. (1966) Solubilities of current purines, pyrimidines and their nucleosides in urea and sucrose solutions. *Bull. Soc. Chim. Belgs.* **75**:673 – 690.

[12]   Herskovits, T.T., Harrington, J.P. (1972) Solution Studies of the Nucleic Acid Bases and Related Model Compounds. Solubility in Aqueous Alcohol and Glycol Solutions. *Biochemistry* **11**:4800 – 4811.
       doi: http://dx.doi.org/10.1021/bi00775a025.

[13]   Herskovits, T.T., Bowen, J.J. (1974) Solution studies of nucleic-acid bases and related model compounds – solubility in aqueous urea and amide solutions. *Biochemistry* **13**:5474 – 5483.
       doi: http://dx.doi.org/10.1021/bi00724a004.

[14]   Scruggs, R.L., Achter, E.K., Ross, P.D. (1972) Thermodynamic effects of exposing nucleic-acid bases to water – solubility measurements in water and organic-solvents. *Biopolymers* **11**:1961 – 1972.
       doi: http://dx.doi.org/10.1002/bip.1972.360110915.

[15]   Camici, M., Sgarrella, F., Ipata, P.L., Mura, U. (1980) The standard Gibbs-energy change of hydrolysis of α-D-ribose-1-phosphate. *Arch. Biochem. Biophys.* **205**:191 – 197.
       doi: http://dx.doi.org/10.1016/0003-9861(80)90098-3.

[16]   Colbert, J.C., Domalski, E.S., Coxon, B. (1987) Enthalpies of combustion of D-ribose and 2-deoxy-D-ribose. *J. Chem. Thermodyn.* **19**:433 – 441.
       doi: http://dx.doi.org/10.1016/0021-9614(87)90128-5.

[17]   Jasra, J.V., Ahluwalia, J.C. (1982) Enthalpies of solution, partial molal heat-capacities and apparent molal volumes of sugars and polyols in water. *J. Solution Chem.* **11**:325 – 338.

[18]   Stern, J.H., Hubler, P.M.(1984) Hydrogen-bonding in aqueous-solutions of D-ribose and 2-deoxy-D-ribose. *J. Phys. Chem.* **88**:1680 – 1681.
       doi: http://dx.doi.org/10.1021/j150653a003.

[19]   Goldberg, R.N., Tewari, Y.B. (1989) Thermodynamic and transport properties of carbohydrates and their monophosphates: the pentoses and hexoses. *J. Phys. Chem. Ref. Data* **18**:809 – 880.
       doi: http://dx.doi.org/10.1063/1.555831.

[20]   Akers, D.L., Goldberg, R.N. (2001) BioEqCalc: A package for performing equilibrium calculations on biochemical reactions. *Mathematica J.* **8**:86 – 113. The package can be downloaded from http://xpdb.nist.gov/enzyme_thermodynamics/.

[21] Miller, S.L., Smith-Magowan, D. (1990) The thermodynamics of the Krebs cycle and related compounds. *J. Phys. Chem. Ref. Data* **19**:1049 – 1073.

[22] Guest, M.F., Pedley, J.B., Horn, M. (1969) Analysis by computer of thermochemical data on boron compounds. *J. Chem. Thermodyn.* **1**:345 – 352.
doi: http://dx.doi.org/10.1016/0021-9614(69)90064-0.

[23] Pedley, J.B. (1972) *Computer Analysis of Thermochemical Data: Catch Tables.* University of Sussex, Brighton, U.K.

[24] Garvin, D., Parker, V.B., Wagman, D.D., Evans, W.H. (1976) *A Combined Least Sums and Least Squares Approach to the Evaluation of Thermodynamic Data Networks.* NBSIR Report No. 76 – 1147. National Bureau of Standards, Washington, D.C.

[25] Alberty, R.A. (2006) *Biochemical Thermodynamics: Applications of Mathematica.* Wiley-Interscience, Hoboken, NJ.

[26] Alberty, R.A. (2007) *Basic Data for Biochemistry.* http://library.wolfram.com/info-center/MathSource/797

[27] Krebs, H.A., Kornberg, H.L., Burton, K. (1957) *A Survey of the Energy Transformations in Living Matter.* Springer-Verlag, Berlin.

[28] Wilhoit, R.C. (1969) *Thermodynamic Properties of Biochemical Substances.* In: *Biochemical Microcalorimetry* (Brown, H.D., Ed). Academic Press, New York; pp. 33 – 81, 305 – 317.

[29] Thauer, R.K., Jungermann, K., Decker, K. (1977) Energy conservation in chemotropic anaerobic bacteria. *Bacter. Rev.* **41**:100 – 183.

[30] Ould-Moulaye, C.B., Dussap, C.G., Gros, J.B. (2001) A consistent set of formation properties of nucleic acid compounds – purines and pyrimidines in the solid state and in aqueous solution. *Thermochimica Acta* **375**:93 – 107.
doi: http://dx.doi.org/10.1016/S0040-6031(01)00522-6.

[31] Ould-Moulaye, C.B., Dussap, C.G., Gros, J.B. (2002) A consistent set of formation properties of nucleic acid compounds – nucleosides, nucleotiodes, and nucleotide-phosphates in aqueous solution. *Thermochimica Acta* **387**:1 – 15.
doi: http://dx.doi.org/10.1016/S0040-6031(01)00814-0.

[32] Benson, S.W., Buss, J.H. (1958) Additivity Rules for the Estimation of Molecular Properties. Thermodynamic Properties. *J. Chem. Phys.* **29**:546 – 572.

[33] Benson, S.W. (1968) *Thermochemical Kinetics.* J. Wiley & Sons, New York.

[34] Benson, S.W., Cruickshank, F.R., Golden, D.M., Haugen, G.R., O'Neal, H.E., Rodgers, A.S., Shaw, R., Walsh, R. (1969) Additivity rules for the estimation of thermochemical properties. *Chem. Rev.* **69**:279–324.
doi: http://dx.doi.org/10.1021/cr60259a002.

[35] Domalski, E.S., Hearing, E.D. (1993) Estimation of the thermodynamic properties of C-H-N-O-S-halogen compounds at 298.15 K. *J. Phys. Chem. Ref. Data* **22**:805–1159.

[36] Domalski, E.S. (1998) Estimation of enthalpies of formation of organic compounds at infinite dilution in water at 298.15 K. In: *Computational Thermochemistry – Prediction and Estimation of Molecular Thermodynamics* (Irikura, K.K., Frurip, D.J., Eds). ACS Symposium Series No. 677, American Chemical Society, Washington, D.C.

BIOGRAPHIES

Karen N. Allen

received her B.S. degree in Biology, *cum laude* from Tufts University and her Ph.D. in Biochemistry from Brandeis University, where she was a Dretzin scholar. Her graduate studies in the laboratory of the mechanistic enzymologist, Dr. Robert H. Abeles, focused on the design, synthesis, and inhibition kinetics of transition-state analogues. Following her desire to see enzymes in action she pursued X-ray crystallography during postdoctoral studies as an American Cancer Society Fellow in the laboratory of Drs. Gregory A. Petsko and Dagmar Ringe. Since 1993 she has lead her own research team at Boston University, first in the Department of Physiology and Biophysics at the School of Medicine, and since 2008 in the Department of Chemistry where she is now a Professor. She is also on the faculty of both the Bioinformatics and Cell and Molecular Biology programs at Boston University.

Dr. Allen's research has focused on the elucidation of enzyme mechanisms and the under-standing of how Nature has evolved new chemistries from existing protein scaffolds. Within this context, her laboratory has plumbed the basis of enzyme-mediated phosphoryl transfer. In addition, Dr. Allen has sought to provide new tools for the exploration of protein structure and function by the invention and implementation of lanthanide binding tags. Dr. Allen's students and postdoctoral researchers have gone on to research positions in structural geno-mics institutes such as RIKEN, Japan and drug discovery companies including AstraZeneca and Novartis as well as in the academic arena as independent investigators. She is Associate Editor of the American Chemical Society journal *Biochemistry*, and has served on both NIH and NSF study sections as a regular panel member. Dr. Allen has been an invited lecturer and seminar speaker on over fifty occasions, and has chaired a number of national and international meetings.

Richard N. Armstrong

Education

1970:	B.S. Western Illinois University, Chemistry, Macomb, Illinois
1975:	Ph.D. Marquette University, Organic Chemistry Milwaukee, Wisconsin (with Prof. N. E. Hoffman)
1976 – 1978:	Postdoctoral Fellow, University of Chicago, Chicago, Illinois (with Prof. E. T. Kaiser)

1970:	B.S. Western Illinois University, Chemistry, Macomb, Illinois
1975:	Ph.D. Marquette University, Organic Chemistry Milwaukee, Wisconsin (with Prof. N. E. Hoffman)
1976 – 1978:	Postdoctoral Fellow, University of Chicago, Chicago, Illinois (with Prof. E. T. Kaiser)
1990:	Sabbatical Leave, Center for Advanced Research in Biotechnology, Rockville, MD, Protein Crystallography (with Gary Gilliland)

Positions Held

since 1997:	Professor, Department of Chemistry, Vanderbilt University, College of Arts & Sciences
since 1995:	Professor, Department of Biochemistry and the Center in Molecular Toxicology, Vanderbilt University School of Medicine
1990 – 1995:	Professor, Department of Chemistry & Biochemistry, UMCP
1988 – 1990:	Chair, Biochemistry Division, UMCP
1985 – 1990:	Associate Professor, Department of Chemistry & Biochemistry, UMCP
1980 – 1985:	Assistant Professor, Department of Chemistry University of Maryland, College Park
1978 – 1980:	Staff Fellow, Laboratory of Bioorganic Chemistry, NIAMDD, National Institutes of Health, Bethesda, Maryland
1976 – 1978:	Postdoctoral Fellow, Department of Chemistry, University of Chicago, Chicago, Illinois

115 publications (refereed articles, book chapters, reviews), more than 120 invited lectures.

Research Interests

Functional genomics. Enzymatic basis of antibiotic resistance. Mechanism and stereochemistry of enzyme-catalyzed reactions. Metabolism and detoxification of drugs and toxic compounds. Protein structure and function and engineering. Protein crystallography. Applications of physical organic chemistry to biochemical and biotechnological problems. Stereochemistry and conformations of strained molecules.

Douglas Auld

received a Ph.D. in Chemistry from University of North Carolina at Chapel Hill followed by post-doc in the Department of Biology at MIT. Prior to joining the NCGC he worked as Assistant Director of Drug Discovery at Pharmacopeia, Princeton, NJ.

Dr. Auld joined the NCGC at its inception and has worked on establishing a center of scientific excellence to serve the research community for assay optimization, HTS and chemical probe distribution. NCGC has developed a concentration-response-based screening approach termed quantitative HTS (qHTS) where compounds are screened at 7 concentrations and applies this to both biochemical and cell-based assays on libraries > 200K in size.

Richard Cammack

is Professor of Biochemistry at King's College, University of London. He graduated with a BA in Biochemistry at the University of Cambridge in 1965 and PhD in Enzymology in 1968, under the supervision of Malcolm Dixon. His research has centred on the use of spectroscopic techniques to study mechanisms of electron transfer and enzyme catalysis, particularly in complex iron-sulfur proteins. He is past Chairman (2000 – 2005) and currently Secretary of the Joint Biochemical Nomenclature committee of IUBMB and IUPAC, and was Editor-in-Chief of the second edition of the Oxford Dictionary of Biochemistry and Molecular Biology. He has published two books, and over 230 research papers. He is currently investigating aspects of the role of iron in health and disease.

María Luz Cárdenas

is a senior research scientist at the CNRS, in the Unit of Bioenergetics and Bioengineering of Proteins in Marseilles. Born in Santiago, she made all her studies at the University of Chile, and became "Doctor en Ciencias" in 1982.

Her doctoral thesis was supervised by Hermann Niemeyer and dealt with the kinetic behaviour of vertebrate "glucokinase", which has the peculiarity of showing positive cooperativity with respect to glucose, despite of being a monomeric enzyme. Her post-doctoral work at the University of Birmingham was done in the laboratory of Ian Trayer in the Biochemistry Department. She acquired British nationality in 1986, having married Athel Cornish-Bowden in 1982.

In 1987, while an Associate Professor of the Faculty of Sciences in the University of Chile she was invited by Jacques Ricard to come to Marseilles, where she joined the CNRS in 1988. In 2002 she was elected as a Corresponding Member of the Chilean Academy of Sciences. Since 2004 she has been a member of the Editorial Board of the Journal of Biosciences. She is an editor of Regard sur la Biochimie, the magazine for members of the French Biochemical Society.She is an enzymologist, a kineticist, and her studies of "glucokinase" led to an invitation from a publishing house in the United States to write a book about this enzyme, and this appeared in 1995. She has about 80 publications in total.

Athel Cornish-Bowden

carried out his undergraduate studies at Oxford, obtaining his doctorate with Jeremy R. Knowles in 1967. After three post-doctoral years in the laboratory of Daniel E. Koshland, Jr., at the University of California, Berkeley, he spent 16 years as Lecturer, and later Senior Lecturer, in the Department of Biochemistry at the University of Birmingham. Since 1987 he has been Directeur de Recherche in three different laboratories of the CNRS at Marseilles. Although he started his career in a department of organic chemistry virtually all of his research has been in biochemistry, with particular reference to enzymes, including pepsin, mammalian hexokinases and enzymes involved in electron transfer in bacteria. He has written several books relating to enzyme kinetics, including Analysis of Enzyme Kinetic Data (Oxford University Press, 1995) and Fundamentals of Enzyme Kinetics (3rd edition, Portland Press, 2004).

Since moving to Marseilles he has been particularly interested in multi-enzyme systems, including the regulation of metabolic pathways. At present his main interest is in the definition of life and the capacity of living organisms for self-organization. In addition his principal areas of research, he has long had an interest in biochemical aspects of evolution, and his semi-popular book in this field, The Pursuit of Perfection, was published by Oxford University Press in 2004.

Lucia Gardossi

is associate professor of organic chemistry at the Faculty of Pharmacy of the University of Trieste.

Since 1987 her research deals with biocatalysis and the application of statistical and computational methods to the study of enzymes. She received the PhD in Medicinal Chemistry in 1995 at the University of Trieste where now she heads the Laboratory of Applied and Computational Biocatalysis. From 1989–1991 she worked on biocatalysis in non-conventional media at MIT in the group of Prof. A.M. Klibanov. Currently she is appointed as Italian delegate inside the scientific board of the European Section of Applied Biocatalysis (ESAB) of the European Federation of Biotechnology and consequently she was involved in the working groups of the European Technology Platform on Sustainable Chemistry.

She worked for two years as researchers also in industry and since 2007 she is the scientific responsible of a spin-off of the University of Trieste focused on biocatalysts development and applications.

Her current scientific interests are on the development of computational methods for the rational development and screening of enzymes and this is also the topic of the EU-Russian project (FP7) she is coordinating at present.

Robert Goldberg

received his Bachelor of Arts (Chemistry Major) from Johns Hopkins University and his Doctor of Philosophy (Physical Chemistry) from Carnegie-Mellon University. After completion of a post-doctoral research fellowship at Mellon Institute and the University of Pittsburgh, he joined the National Institute of Standards and Technology (formerly the National Bureau of Standards) in 1969. His primary areas of expertise include chemical thermodynamics, calorimetry and equilibrium measurements, data evaluation, thermodynamics of solutions, biochemical thermodynamics, analytical microcalorimetry, and chromatography. A major focus of his research has been on the thermodynamics of enzyme-catalyzed reactions. This has resulted in the determination of the thermodynamic parameters for a large number of such reactions – including a substantial number of the most important reactions pertinent to physiology and to metabolism as well as reactions that are of major industrial interest. These studies involve the combined use of equilibrium and calorimetric measurements coupled with thermodynamic modeling calculations. The information obtained allows for the prediction of the position of equilibrium of the studied reaction(s) over wide ranges of temperature, pH, and ionic strength.

Recent research has included studies of biochemical reactions in non-aqueous solvents, redox reactions, and reactions in the shikimate and chorismate pathways.

Codes for performing equilibrium calculations on systems of biochemical reactions have also been developed and published. The entire field involving the thermodynamics of enzyme-catalyzed reactions has been surveyed and the data extracted and made available on the web: http://xpdb.nist.gov/enzyme_thermodynamics/. He has been active in IUPAC, ASTM, and in the U.S. Calorimetry Conference. He is a recipient of the Measurement Services Award of the National Institute of Standards and Technology. He is presently a scientist emeritus at the National Institute of Standards and Technology.

Carsten Kettner

studied biology at the University of Bonn and obtained his diploma at the University of Göttingen in the group of Prof. Gradmann which had the pioneering and futuristic name – "Molecular Electrobiology". This group consisted of people carrying out research in electrophysiology and molecular biology in fruitful cooperation. In this mixed environment, he studied transport characteristics of the yeast plasma membrane using patch clamp techniques. In 1996 he joined the group of Dr. Adam Bertl at the University of Karlsruhe and undertook research on another yeast membrane type. During this period, he successfully narrowed the gap between the biochemical and genetic properties, and the biophysical comprehension of the vacuolar proton-translocating ATP-hydrolase. He was awarded his Ph.D for this work in 1999. As a post-doctoral student he continued both the studies on the biophysical properties of the pump and investigated the kinetics and regulation of the dominant plasma membrane potassium channel (TOK1). In 2000 he moved to the Beilstein-Institut to represent the biological section of the funding department. Here, he is responsible for the organization of symposia (Beilstein "Bozen" Symposium, Glyco-Bioinformatics and ESCEC), research (proposals). Since 2004 he coordinates the work of the STRENDA commission and promotes along with the commissioners the proposed standards of reporting enzyme data. In 2007 he became involved in the invention of a program for the establishment of Beilstein Endowed Chairs for Chemical sciences and related sciences.

Scott T. Lefurgy

is a postdoctoral fellow at the Albert Einstein College of Medicine in the laboratory of Tom Leyh. He received his Ph.D. in the laboratory of Virginia Cornish at Columbia University, where he studied protein function using in vivo screening and selection methods. He was the recipient of a National Defense Science & Engineering Graduate Fellowship and a Barry Goldwater Scholarship. Scott holds Bachelor of Science (Biochemistry) and Music (Voice Performance) degrees from the University of Michigan. An accomplished classical singer, he made his Carnegie Hall debut in 2003 and was favorably reviewed by the international magazine Opera in 2005.

Klaus Mauch

is co-founder and CEO of Insilico Biotechnolgy. In his previous role of the company's CTO, he was responsible for the design and development of Insilico's modeling and simulation platform by establishing novel methods for computer aided construction and analysis of cellular networks. As a group leader at the Institute of Biochemical Engineering at the University of Stuttgart, Klaus Mauch gained extensive experience in metabolic engineering. He received a Masters in Chemical Engineering from the Technical University of Karlsruhe, Germany.

Frank Raushel

Education

| 1972 | B.A.: St. Thomas University, St. Paul, MN chemistry (*magna cum laude*) |
| 1976 | Ph.D.: University of Wisconsin-Madison |

Postdoctoral Studies

1976 – 1980	The Pennsylvania State University

Professional Appointments

2004 -	Davidson Professor of Science
1989 -	Professor of Chemistry, Department of Chemistry, Biochemistry & Biophysics, Texas A&M University
2008 -	Professor of Chemistry, Department of Chemistry, Biochemistry & Biophysics, Texas A&M University
2008	Visiting Professor, Universität Regensburg, Institut für Biophysik und Physikalische Biochemie
1992 – 1993	Visiting Professor, Enzyme Institute, University of Wisconsin
1986 – 1989	Associate Professor, Department of Chemistry, Department of Biochemistry
1980 – 1986	Assistant Professor, Department of Chemistry, Texas A&M University
1976 – 1980	Research Associate, Department of Chemistry, Penn State University

Honors

2009	Repligen Award (American Chemical Society)
2000	TAMU Association of Former Students Award for Research

Research Interests

Elucidation of enzyme reaction mechanisms and protein structure using kinetic, magnetic resonance, X-ray crystallography, and genetic techniques.

Johann Rohwer

is Associate Professor in the Department of Biochemistry at Stellenbosch University, South Africa. He obtained his Ph.D. in 1997 from the University of Amsterdam, working on the control and regulation of the bacterial phosphotransferase system under the supervision of Hans Westerhoff. He then joined Stellenbosch University, where he and his colleagues Jannie Hofmeyr and Jacky Snoep constitute the "Triple-J Group for Molecular Cell Physiology", a research group that studies the control and regulation of cellular processes using theoretical, numerical and experimental approaches.

Johann has contributed to the theoretical development of metabolic control analysis, to its experimental application, and to the development of software tools for computational systems biology. His main research interests are the construction of kinetic models of cellular function with a particular emphasis on plant central carbon metabolism, and the application of NMR spectroscopy to the non-invasive study of metabolism *in vivo*. He has received the President's Award from the South African National Research Foundation and the Silver Medal of the South African Society of Biochemistry and Molecular Biology (SASBMB). Together with the other Triple-Js, he has chaired the BTK: International Study Group for Systems Biology, and has represented his university on the South African National Bioinformatics Network. In 2008, he was the recipient of an Alexander von Humboldt research fellowship to collaborate with Mark Stitt at the Max Planck Institute of Molecular Plant Physiology in Potsdam, Germany. Johann is an Associate Editor of BMC Systems Biology and the Secretary of the SASBMB.

Hartmut Schlüter

Institution and Position:

University Hospital Hamburg-Eppendorf
Full Professor

1988:	Diploma (=M.Sc.) in Biochemistry, Faculty of Chemistry, University of Münster
1991:	Ph.D. (Dr. rer. nat.) in Biochemistry, University of Münster, Faculty of Chemistry, Thesis supervisor: Prof. Dr. H. Witzel
1994:	Heinz Maier-Leibnitz prize
1995:	Gerhard Hess award (DFG)
1995:	Bennigsen-Foerder prize
1991 – 1996:	Postdoctoral fellowship at the Medical Faculty of the University of Münster
1996:	Habilitation (Dr. rer. nat. habil.) in Pathobiochemistry at the Medical Faculty of the University of Münster
1996 – 2000:	Group leader at the Medical Faculty of the Ruhr-University of Bochum
2000 – 2008:	Senior Scientist and Head of the Bioanalytical Laboratory of Nephrology, Charité – University Medicine Berlin, Campus Benjamin-Franklin, Joint Facility of the Free University of Berlin and the Humboldt-University of Berlin
2003 – 2008:	Professor at the Charité, Campus Benjamin-Franklin, University-Medicine Berlin

2003:	Member of the board of the Center for Functional Genomics – Berlin-Brandenburg (http://www.cffg.de/)
2004 – 2008:	Head of the Core-Facility Protein Purification
2005:	Founding member of the Mass-Spec-Net Berlin-Brandenburg (http://www.massspecnet.de/)
2007:	Appointment to a professorship for Mass Spectrometric Proteomics at the University Hospital Hamburg-Eppendorf.
2008-current:	Full professor for Mass Spectrometric Proteomics at the University Hospital Hamburg-Eppendorf

Working field(s):

Biochemistry and physiology of diadenosine polyphosphates
Desease-associated proteases
Functional Proteomics – Enzymology
Liquid chromatography of biomolecules
Mass spectrometry of biomolecules
Protein purification

Dietmar Schomburg

1974:	Diplom in Chemistry at the Technical University "Carolo-Wilhelmina" in Braunschweig
1976:	Dr. rer. nat. in Chemistry (Structural Chemistry of Organo-phosphorus compounds)
1985:	Habilitation (Dr. rer. nat. habil.) for Structural Chemistry

Scientific Career:

1976 – 1978:	Post-Doc in the Chemistry Department at Technical University Braunschweig.
1978 – 1979:	Research Fellow at Harvard University in Cambridge, Mass., U. S. A. in Professor W.N. Lipscomb's and Professor F.H. Westheimer's groups.
1979 – 1981:	Post-Doctoral Fellow in the Chemistry Department at Braunschweig Technical University
1981 – 1983:	Assistant Professor (Hochschulassistent), Braunschweig Technical University

1983 – 1986:	Head of the X-ray lab at the German Centre for Biotechnology – GBF (Gesellschaft für Biotechnologische Forschung), Braunschweig
1987 – 1996:	Head of the GBF Department of "Molecular Structure Research."
1989 – 1995:	Head of CAPE (Center of Applied Protein Engineering)
1990 – 1996:	(apl.) Professor at the Technical University Braunschweig
1996 – 2007:	Full Professor of Biochemistry, University of Cologne
since 2007:	Full Professor of bioinformatics & biochemistry, Technical University Braunschweig

Christoph Steinbeck

was born in Neuwied, Germany, in 1966. He studied Chemistry at the University of Bonn, where he received his diploma and doctoral degree in the workgroup of Prof. Eberhard Breitmaier at the Institute of Organic Chemistry. Focus of his Ph. D. thesis was the program LUCY for computer assisted structure elucidation. In 1996, he joined the group of Prof. Clemens Richert at Tufts University in Boston, MA, USA, where he worked in the area of biomolecular NMR on the 3D structure elucidation of peptidenucleic acid conjugates.

In 1997 Christoph Steinbeck became head of the Structural Chemo and Bioinformatics Workgroup at the newly founded Max Planck Institute of Chemical Ecology in Jena, Germany. In Fall 2002 he moved to Cologne University Bioinformatics Center (CUBIC) as head of the Research Group for Molecular Informatics. His research focuses on methods for Computer Assisted Structure Elucidation in Metabolomics and Natural Products Research.

In December 2003 Christoph Steinbeck received his Habilitation in Organic Chemistry from Friedrich Schiller University in Jena, Germany. His group develops a number of the leading open source software packages in Chemo and Bioinformatics, including the Chemistry Development Kit (CDK), a Java library for chemo and bioinformatics, NMRShiftDB, an open content database for chemical structures and their NMR data, and Bioclipse, an Eclipse based Rich Client for everything and nothing in particular in molecular informatics. Dr. Steinbeck is chairman of the Computers Information Chemistry (CIC) division of the German Chemical Society, trustee of the Chemical Structure Association (CSA) Trust, and member of various editorial boards and committees.

Today, Christoph Steinbeck is Head of Chemoinformatics and Metabolism at the European Bioinformatics Institute (EBI) in Hinxton, Cambridge, UK.

Christopher Taylor

Chris Taylor works at the European Bioinformatics Institute from where he coordinates the MIBBI project (mibbi.org) with colleagues. The MIBBI project offers access to 'Minimum Information' reporting guidelines for diverse bioscience domains. Chris also has leading roles in several domain-specific standardisation projects, such as the HUPO Proteomics Standards Initiative, and participates in many more. His largely redundant doctorate in population genetics and speciation was conferred in Manchester, UK.

Keith Tipton

Degrees etc.

B.Sc. (Biochemistry), St Andrews University (1962); M.A. (1965), Ph.D. (1966); Cambridge University; M.R.I.A. (1984)

Main Posts:

1965 – 1977:	University of Cambridge: Demonstrator & Lecturer;
1965 – 1977:	Fellow of King's College Cambridge;
1997 – present	University of Dublin: Professor of Biochemistry;
1979- present	Fellow of Trinity College, Dublin;
1976, 1993 & 2003	Visiting Professor: Universities of Florence & Siena (1987 & 1999);
1988 – 89	Autonomous University of Barcelona

Publications:
Over 250 papers in refereed journals; 35 papers as chapters in books; editor of 19 books, > 150 abstracts; 1 patent, co-author of three books.

Research Interests:
Enzymology: regulation, kinetics, inhibition, isolation, applications and classification. Metabolic analysis and simulation. Neurochemistry: depression, degenerative diseases and 'neuroprotection'. Biochemical Pharmacology: drug design, ethanol.

Karen van Eunen

2009 – present	Post-doctoral researcher in Food Science at the Department of Chemical and Biological Engineering, Chalmers University of Technology, Gothenburg, Sweden.
Supervisor:	Dr. Thomas Andlid. Project: Folate metabolism in Yeast.
2004 – 2009	PhD student at the Dept. Molecular Cell Physiology, Faculty of Earth and Life Sciences, Vrije Universteit Amsterdam. Promotor: Prof. Dr. Hans V. Westerhoff and Co-promotor: Dr. Barbara M. Bakker. The project dealt with (i) how the flux through yeast glycolysis is regulated upon nitrogen starvation (ii) development of an assay medium to measure enzyme capacities under *in vivo*-like conditions and (iii) measure protein and synthesis rates of specific proteins. This project is part of the project: 'Vertical Genomics from gene expression to function and back'.
2002 – 2004	Scientist of the Research and Development department of Heineken Nederlands Beheer BV. Worked on EU-project: Stress-tolerant industrial yeast strains for High-Gravity Brewing.
1999 – 2001	M. Sc. Biology (*Cum Laude*), Leiden University
1995 – 1999	B. Sc. Medical Biotechnology, Hoger Laboratorium Onderwijs, Utrecht, Hogeschool van Utrecht.

Yu (Brandon) Xia

received his B.S. in Chemistry (major) and Computer Science (minor) from Peking University, and his Ph.D. in Chemistry from Stanford University. While at Stanford, he worked on computational structural biology as a Howard Hughes Medical Institute Pre-doctoral Fellow. Following that he worked on protein bioinformatics as a Jane Coffin Childs Fellow at Yale University. He is currently an Assistant Professor in the Bioinformatics Program and the Department of Chemistry at Boston University, with a secondary appointment in the Department of Biomedical Engineering. He has published over 35 research articles, scientific reviews, and book chapters. His research interests include the prediction and analysis of protein structures and networks.

Index of Authors

A

Allen, Karen N.   67
Apweiler, Rolf   79
Armstrong, Richard N.   1
Auld, Douglas S.   21

B

Bakker, Barbara M.   165
Boxer, Matthew B.   21
Brown, Daniel W.   1

C

Cammack, Richard   87
Cárdenas, María Luz   107
Cook, Paul D.   1
Cornish-Bowden, Athel   107
Curien, Gilles   107

D

Dumas, Renaud   107
Dunaway-Mariano, Debra   67

F

Franzosa, Eric A.   99

G

Gardossi, Lucia   187
Goldberg, Robert N.   213

H

Halling, Peter J.   187
Hofmann, Ute   123
Hofmeyr, Jan-Hendrik S.   149
Holzhütter, Hermann-Georg   79

I

Inglese, James   21

J

Jaster, Robert   137
Jungblut, Peter R.   79

K

Kengne, Sylvestre   205
Kiewiet, José   165

L

Lefurgy, Scott T.   45
Leyh, Thomas S.   45, 59
Lynagh, Kevin J.   99

M

Maier, Klaus   123
Mauch, Klaus   123
Müller, Wolfgang   205

R

Rateitschak, Katja   137
Raushel, Frank M.   9
Reuss, Matthias   123
Rohwer, Johann M.   149
Rojas, Isabel   205

S

Sauer-Danzwith, Heidrun   205
Schlüter, Hartmut   79
Shen, Min   21
Southall, Noel   21

T

Taylor, Chris F.   181
Thomas, Craig J.   21
Thorne, Natasha   21

V

van Eunen, Karen   165

W

Weidemann, Andreas   205
Westerhoff, Hans V.   165
Wittig, Ulrike   205
Wolkenhauer, Olaf   137

X

Xia, Yu   99

245

Index   Experimental Standard Conditions of Enzyme Characterizations, September 13th – 16th, 2009, Rüdesheim/Rhein, Germany

Index

246

Experimental Standard Conditions of Enzyme Characterizations, September 13th – 16th, 2009, Rüdesheim/Rhein, Germany **Index**

248

Experimental Standard Conditions of Enzyme Characterizations, September 13th – 16th, 2009, Rüdesheim/Rhein, Germany **Index**

249

Index   Experimental Standard Conditions of Enzyme Characterizations, September 13th – 16th, 2009, Rüdesheim/Rhein, Germany